T0191817

Environmental Footprints and Eco-design of Products and Processes

Series Editor

Subramanian Senthilkannan Muthu, Head of Sustainability - SgT Group and API, Hong Kong, Kowloon, Hong Kong

Indexed by Scopus

This series aims to broadly cover all the aspects related to environmental assessment of products, development of environmental and ecological indicators and eco-design of various products and processes. Below are the areas fall under the aims and scope of this series, but not limited to: Environmental Life Cycle Assessment; Social Life Cycle Assessment; Organizational and Product Carbon Footprints; Ecological, Energy and Water Footprints; Life cycle costing; Environmental and sustainable indicators; Environmental impact assessment methods and tools; Eco-design (sustainable design) aspects and tools; Biodegradation studies; Recycling; Solid waste management; Environmental and social audits; Green Purchasing and tools; Product environmental footprints; Environmental management standards and regulations; Eco-labels; Green Claims and green washing; Assessment of sustainability aspects.

More information about this series at http://www.springer.com/series/13340

Eric Lichtfouse ·
Subramanian Senthilkannan Muthu · Ali Khadir
Editors

Inorganic-Organic Composites for Water and Wastewater Treatment

Volume 1

Editors
Eric Lichtfouse
Aix-Marseille University
Marseille, France

Subramanian Senthilkannan Muthu
SgT Group & API
Hong Kong, Kowloon, Hong Kong

Ali Khadir
Islamic Azad University of Shahre Rey Branch
Teheran, Iran

ISSN 2345-7651 ISSN 2345-766X (electronic)
Environmental Footprints and Eco-design of Products and Processes
ISBN 978-981-16-5918-8 ISBN 978-981-16-5916-4 (eBook)
https://doi.org/10.1007/978-981-16-5916-4

This Springer imprint is published by the registered company Springer Nature Singapore Pte Ltd.
The registered company address is: 152 Beach Road, #21-01/04 Gateway East, Singapore 189721, Singapore

Contents

About the Editors

Eric Lichtfouse is a research scientist at Aix-Marseille University and an invited professor at Xi'an Jiaotong University. His research interests include climate change, carbon, pollution and organic compounds in air, water, soils and sediments. He is teaching biogeochemistry and scientific writing. He is chief editor of the journal Environmental Chemistry Letters and the book series Sustainable Agriculture Reviews and Environmental Chemistry for a Sustainable World.

Subramanian Senthilkannan Muthu currently works for SgT Group as Head of Sustainability and is based out of Hong Kong. He earned his PhD from The Hong Kong Polytechnic University and is a renowned expert in the areas of environmental sustainability in textiles & clothing supply chain, product life cycle assessment (LCA) and product carbon footprint assessment (PCF) in various industrial sectors. He has five years of industrial experience in textile manufacturing, research and development and textile testing and over a decade of experience in life cycle assessment (LCA), carbon and ecological footprint assessment of various consumer products. He has published more than 100 research publications, written numerous book chapters and authored/edited over 100 books in the areas of carbon footprint, recycling, environmental assessment and environmental sustainability.

Ali Khadir is an environmental engineer and a member of the Young Researcher and Elite Club, Islamic Azad University of Shahre Rey Branch, Tehran, Iran. He has published several articles and book chapters in reputed international publishers, including Elsevier, Springer, Taylor & Francis and Wiley. His articles have been published in journals with IF greater than 4, including the Journal of Environmental Chemical Engineering, International Journal of Biological Macromolecules and Journal of Water Process Engineering. He has been also the reviewer of journals and international conferences. His research interests center on emerging pollutants, dyes and pharmaceuticals in aquatic media, advanced water and wastewater remediation techniques and technology.

Wastewater Treatment Technologies

Oluwaseun Jacob Ajala, Jimoh Oladejo Tijani, Mercy Temitope Bankole, and Ambali Saka Abdulkareem

Abstract Access to sustainable drinking water remains a serious challenge both in rural and urban areas due to population explosion, pollution of various water systems through natural and anthropogenic activities such as industrial activities, agricultural practices, urbanisation among others. Although, water and wastewater treatment was early discovered in history, however, took several years before people understood that adequate judgement for quality water is not by human sense. The earliest water treatment includes filtration and removal of pathogens carriers of typhoid, cholera and dysentery among other diseases. Different treatment techniques such as membranes filtration technologies, thermal technologies, biological aerated filter technologies, advanced oxidation technologies are available for the mitigation of organic, inorganic, heavy metals, microorganism, suspended solids and emerging persistent pollutants from water. This chapter review different technological of advanced oxidation processes wastewater technologies especially their efficiencies, cost implications, durability among others. Of all the advanced oxidation processes, photoelectrocatalysis has been found more suitable for the treatment of wastewater than conventional methods.

Keywords Physico-chemical parameters · Classification · Conventional methods · Advanced oxidation processes · Photoelectrocatalysis · Photo-fenton · Photocatalysis

---

O. J. Ajala (✉) · J. O. Tijani · M. T. Bankole · A. S. Abdulkareem
Department of Chemistry, Federal University of Technology Minna, Minna P. M. B. 65,, Bosso State, Nigeria

J. O. Tijani
e-mail: jimohtijani@futminna.edu.ng

M. T. Bankole
e-mail: bankole.temitope@futminna.edu.ng

A. S. Abdulkareem
e-mail: kasaka2003@futminna.edu.ng

Nanotechnology Research Group, Africa Center of Excellence for Mycotoxin and Food Safety, Federal University of Technology, Minna P. M. B. 65,, Niger State, Nigeria

E. Lichtfouse et al. (eds.), *Inorganic-Organic Composites for Water and Wastewater Treatment*, Environmental Footprints and Eco-design of Products and Processes,
https://doi.org/10.1007/978-981-16-5916-4_1

1 Introduction

Despite the importance of water to human life especially for drinking, household chores, agricultural and industrial purposes, the quality of water has been compromised through the release of different pollutants into water system [127]. Some of these pollutants are release either through domestic waste or industrial waste or both. Some of these domestic and industrial wastes are highly toxic to living beings and our ecosystem. They have caused our water systems to experience different colours, high turbidity, bad odours, suspended solid matters, heavy metals, inorganic and organic and micro-organisms [125] which led to water pollution.

In 2019, United Nations reported that water pollution has cause loss of many lives more than any other form of violence dues to several diseases such as water borne diseases (i.e. Diarrhea), type B and type C hepatitis virus and human Immune deficiency virus among others. One of the most significant disease's burdens among many groups of lives is waterborne diseases, especially in developing countries. In Nigeria, the ratio of those that suffered from diarrhea is 1:25 among children below age five [126]. Thus, poor management of solid and liquid waste has resulted to spread of many diseases [101].

Also, in 2007 World Health Organization estimated that yearly patients between 8 to 16 million type B hepatitis virus, 2.3 to 4.7 million type C hepatitis virus and 80,000 to 160,000 HIV (human immune deficiency virus) always exist which all result from poor waste management systems [117]. WHO further estimated that annual death of approximately 829,000 which translate to 54–65% caused by diarrhea in low- and middle-income countries are due to consumption of unconventional water [90]. It was discovered that around 13% of acute infections which translated to about 370,000 deaths in 2016 was linked to inadequate access to safe drinking water [90]. The use of sewage wastewater in agriculture for irrigation may also result in disease transmission [126]. In 2016, malaria cases reported were 217 million with over 451, 000 deaths equivalent to 90% in sub-Saharan Africa alone [128]. World Health Organization [126] reported that the population of the world is expected to grow by 33% (7.2 to 9.6 billion) between 2014 and 2050 with increase in urban population by 61% (3.9–6.3 billion) in Africa and Asia. Climate variability and climate change are also pilling pressure on water supplies due to drought and flooding [27]. Due to the problem of water pollution and water shortage, there is a need to retreat, reuse and recycle wastewater obtained via different wastewater treatments.

2 Physico-Chemical Assessment of Wastewater

In this section, physico-chemical parameters used to evaluate the quality of water are discussed. This will help the reader to understand some important parameters to known if water was polluted. Some of the physical and chemical parameters used to determine the quality of wastewater are explained as follows.

2.1 Physical Characteristics

(I) Colour

Colour is one of the essential water quality parameters that indicates the period of time of exposure. Generally, fresh wastewater always appeared grey while after a long period, the wastewater is always black due to the decomposition of bacterial via anaerobic conditions [17]. It has been found that blacken colouration of wastewater, normally occurs since there is generation of various sulphides such as ferrous sulphide or hydrogen sulphide via reaction of with divalent metals such as iron under anaerobic conditions [138].

(II) Odour

Odour determination is becoming more important, as the public is becoming more concerned. Odour is usually produced via gases during decomposition process of the organic matter [132]. There is distinctive but not offensive for fresh wastewater which is less objectionable compare with septic wastewater odour dues to several odorous compounds release during biological decomposition through anaerobic conditions [32].

The sources of different unpleasant odours generated by industrial wastewater are identified in the Table 1.

(III) Temperature

Wastewater temperature is always greater than the temperature of fresh water dues to the activities of aerobic bacteria which lower the presence of oxygen thereby increase the water temperature [109]. Temperature of wastewater is very important since there are some treatment schemes for wastewater such as biological processes which depends on temperature [15]. Though there are some factors causing variation in temperature such as seasons, geographic location among others, it has been observed that wastewater temperature is always higher than the fresh water [136]. In the Equatorial regions, its temperature varies from 13 to 24 °C while in Polar region, temperature of wastewater varies from 7 to 18 °C [82].

Table 1 Odour sources in selected industries wastewaters

Industries	Sources of odours
Cement work	Dibutyl sulphide Amines, Mercaptans, SO_2, H_2S
Food	Fermentation produces, sulphides, amines, mercaptans
Rubber	Mercaptans, sulphides
Paper pulp	Sulphur dioxide, hydrogen sulphide
Pharmaceutical	Fermentation produces
Textile	Phenolic compounds
Organics compost	Sulphur, ammonia compounds

(IV) Total Solids

Wastewater total solid comprises soluble compounds and suspended or insoluble solids dissolved in water. The content of the insoluble solid is the dried residue removed through filtration process [116]. The volatile solids burned off as a result of ignition of residue. Several organic matters are volatile solids in nature, though some organic matter cannot be ignited while at high temperature, some inorganic salts break down [58]. The following are the organic matters; fat, carbohydrates and proteins which constituted around 40–65% of the solids in the average suspended wastewater. Settled solids are those solids that can be removed by sedimentation [105]. 60% of the suspended solid in municipal wastewater are settled solid which are been expressed as Millilitres per litre [9].

2.2 Chemical Characteristics

(I) Phosphorus and Nitrogen

These elements can be determined through the measurement of phosphorus concentration and nitrogen concentration present in the water respectively [81]. One of life essential nutrients is Phosphorus, which can be derived from fertilizers, animal waste and phosphate detergents. When the concentration of phosphorus is high, orthophosphate will affect the quality of the water by increasing algae growth and further lower the volume of oxygen needed for the aquatic animals and eventually led to their death whereby pollute the water [129]. Orthophosphate will also cause odours and tastes in the water.

Nitrogen is another essential nutrient that is very important to plants in ecosystem [134]. The major sources of nitrogen include; animal waste and fertilizer while the other sources are; atmospheric deposition and septic systems. When application of nitrogen such as through fertilizers is higher than the amount that can be ingested by the plants or released into the atmosphere via denitrification, then the nitrogen concentration will increase [67]. When nitrate is in excess, it is not toxic to aquatic life though increase in nitrogen concentration may lead to overgrowth in algae which reduce the content of oxygen of the water thereby kills or damage aquatic species such as fish and pollute the water [74].

(II) Chemical Oxygen Demand

The chemical oxygen demand (COD) is an essential parameter used to determine the water quality through the measurement of the amount of oxygen to oxidize particulate and soluble organic matter [118]. Similar to biological oxygen demand (BOD), COD provides an index to assess the effect of wastewater discharged. When the COD level is high, the level of dissolved oxygen reduces and can lead to anaerobic conditions and deteriorate the aquatic life [68]. It can be measured using COD analyzers and it has a shorter length of testing time compared to Biological Oxygen Demand.

(III) Dissolved Oxygen

This is the determination of the amount of oxygen present in a sample of water or wastewater at the time of collection and it is very important for respiration of aquatic life such as fish and other organisms [68]. This dissolved oxygen get access into the water system through by- product derived from photosynthesis of plants and from atmosphere diffusion. The dissolved oxygen concentration in the epilimnion of the waters bodies continue to equilibrate with concentration of atmospheric oxygen as it is 100% dissolved oxygen saturated [37]. The oxygen diffusion rate to the atmosphere is lower than the photosynthesis rate when the algae growth is in excess and over saturates the water with dissolved oxygen. When the dissolved oxygen is low there is no mechanism to replace the consumed oxygen via decomposition and respiration which will lead to death of aquatic animals [43].

(IV) Biological Oxygen Demand

This is referred to as the estimation of organic matter amount that is unstable in the water. It estimates the required amount of oxygen needed by the microorganism to degrade the available organic matter into simplest forms such as carbon (IV) oxide, water and ammonia [5]. Though it helps to establish unstable organic matter characterization, it takes a much longer period of time compared to chemical oxygen demand.

(V) pH

The potential hydrogen (pH) is the measurement of the basicity or acidity of a water body [79]. The scale of pH is in logarithm form which commonly ranges from 14 to 0. When the pH value of a solution is 6, it is ten times more acidic than a solution with pH value of 7 [108]. The pH value of sachet water is 7. which is neutral, when the pH value of Water is said to be less than 7, it is considered to be acidic and when the pH value of water is said to be greater than 7, it is considered to be alkaline or basic [3].

(VI) Conductivity

Conductivity of water is an expression of its capacity to transfer electric current in an aqueous solution [43]. It is estimated in micro siemens per centimetre (μs/cm) or micromhos per centimetre (μmhos/cm). The conductivity of distilled water is said in the range of 0.5–3 μs/cm. This ability depends on the mobility, presence of ions, valence, their total concentration, relative concentrations and on the temperature of the liquid [43]. The following solutions are relatively good conductors such as salts, bases and inorganic acids among others.

(VII) Alkalinity

Alkalinity is a measurement of the total components present in the water that tend to neutralize acidity [43]. This is carried out by determining various levels of radicals. Such radicals are hydroxides, phosphates, bicarbonate and carbonates present in the water which is measured in terms of milligram per litre of calcium, carbonate (that is ppm of $CaCO_3$). This also determined the ability of water to resist the change in pH. The organisms are being affected directly due to the change in the pH values caused by the presence of some pollutant toxicity in the water system [95].

3 Classification of Pollutants in Wastewater

There are two classes of pollutants to be discussed in this section; they are: Organic Pollutants and Inorganic Pollutants.

3.1 Organic Pollutants

(I) Dyes

Generally, dyes are organic compounds which been classified into cationic, anionic and nonionic dyes. These dyes are highly toxic and potentially carcinogenic which are related to various diseases in humans and animals [147]. Dyes are also related to environmental degradation such as recalcitrant in aerobic environments [8] whereby responsible for its bioaccumulation which further entered the water system [63]. There are some of these dyes that degraded partially in the presence of anoxic sediments like in the case of azo-type compounds which occurs during reduction and further release dangerous aromatic amines [63] and in some cases, the dyes undergo direct combination with intermediate substance to produced carcinogenic or mutagenic compounds [63]. The effects of recalcitrant and xenobiotic nature of the dyes end on impacting the functioning and structure of the ecosystem [53]. In long period exposures, some dyes with a longer half life of 2–13 years always bring profound unfolding such as to human health or to aquatic biota as it is with metal-complex dyes [94].

(II) Pesticide and Pesticides Residue

Pesticides are categories as herbicides, molluscicides, fungicides, rodenticides, nematocides insecticides and plant growth among others. Globally, it has been widely used to control the pests which cause various diseases and increase the yield of crops yield. However, according to several studies, pesticide has been proved to inflict adverse effects on human health and our environment [4].

These pesticides are designed and produced with the view to kill insect pests generally and not site-specific. The applications designed of these pesticides are expected to operate in contact or systemic approach with the target pest in order to kill and to avoid the non target living things. However, some of these target pests are simply species animals that have similar characteristics to other animals such as susceptibility to certain toxins. Therefore, a chemical that is toxic to an animal may also be toxic to other animals. It has been reported by some researchers that larger dose of some pesticides can kill humans while the normal does require to kill some pests can further affect human health such as sex hormones disruption and lowering of reproductory system performance among others [104]. These pesticides can act as a xeno-hormones whereby mimicked the action of endogenous hormones, hence it has been proved and categorized as endocrine disruptors [10].

(III) Pharmaceutical Pollutants

Pharmaceuticals and their metabolites have been generating serious issues in the ecosystem especially in the aquatic environment and have become a major concern to researchers in the area of environment chemistry [69]. Till date, majority of articles published to address the presence of drugs in the effluent of the sewage. Exposures to these drugs are associated with some risks which are significant to the natural environment [98]. In countries where reuse of water is practice, public's concern is more focused on exposure to human health [20].

(IV) Volatile Organic Carbons

The following are the examples of volatile organic compounds: toluene, benzene, xylenes, trichloroethylene, trichloroethane and dichloromethane are common pollutants present in soil [143]. In commercialized areas, one of these sources of these pollutants is leakage of underground storage tanks. Improper release of solvents and landfills are also significant sources of volatile organic carbon in the soil [6]. Some of these organic substances are classified as priority pollutants namely; polycyclic aromatic, polychlorinated biphenyls, formaldehyde, acetaldehyde, 1, 2-dichloroethane, dichloromethane 1, 3-butadiene and hexachlorobenzene among others [46].

Several industrial wastewaters generated from industries contain heavy metals. For instance, electroplating industries produce qualities of heavy metals in its wastewater. The following are these heavy metals released during their operations and application; zinc, chromium, copper, platinum, cadmium, lead, nickel, vanadium, silver and titanium among others [131]. This application includes; milling, electroplating, anodizing-cleaning conversion-coating and etching [131]. Another source of these heavy metals is the wastes of printed circuit board industry. Nickel, Tin and Lead solder plates are commonly used resistant over plates [100], the products of inorganic pigment industry which is pigments contain cadmium sulphide and chromium molecules [135]. All these generated large quantities of wastewaters with residues and hazardous sludges [122].

3.2 *Inorganic Pollutants*

(I) Anionic Pollutants

Fluoride, chromium and arsenic are the most commonly found anionic pollutants in wastewater. Fluoride is one of the species present in aqueous media. In Arsenic species, there are Arsenate (As(III)) and Arsenate (As(V)) which were more toxic than organic matters [103]. Arsenate is said to be predominately much on the surface of water while Arsenite is dominant in ground water system [61]. Arsenite is in negative ionic form ($H_2AsO_3^-$) under the condition of pH while at pH value of 2.2 As(V) is present as H_3AsO_4, at pH value of 2.2–6.98 As(V) is present as $H_2AsO_4^-$, at pH value of 6.98–11.5 As(V) is present as $HAsO_4^{2-}$ and at pH 11.5 As(V) is present as AsO_4^{3-} [51]. Chromium is of two different forms in the environment; Cr(III) and Cr(VI) with the latter considered more toxic than the former [31]. The predominant species of Cr(VI) is $HCrO_4^-$ at low pH but when the pH value increases, there is a shift to CrO_4^{2-} or $Cr_2O_7^{2-}$. However, in anionic pollutants, the surface charged is negative as the solution pH value increases which weakening the forces of electrostatic attraction between negatively charged anionic pollutants and the adsorbent, thus reduced the efficiency of the adsorbent [92]. Moreover, when there is an increase in pH value, the anionic pollutants compete with OH^- especially at level of higher pH.

(II) Cationic Pollutants

Co(II), Ni(II), Cu(II), Cd(II), Pb(II), Hg(II) and Zn(II) are the generally known cationic pollutants present in wastewater [56]. Any cation with atomic mass greater than 23 is referred to as heavy metal [120]. Several of these heavy metals such as Mercury, Lead, Chromium and Cadmium among others are dangerous to human health and ecosystem. Lead and Zinc can lead to corrosion and are harmful to humans. Heavy metal cannot deteriorate unlike organic pollutants, thus it is difficult to treat via remediation.

There is three mechanisms of adsorption which determined metal ion adsorption. They are ion exchange, electrostatic and complex formation [45]. Graphene oxide sheets have abundant oxygen groups which can be used to bind ions via ion exchange, electrostatic and coordinate approaches. For those cationic pollutants, when the pH value is high it easy adsorption process dues to the competition of –O– and COO– on the sites between metal ions and proton in acidic conditions which lower adsorption capacity. When there is increased in the solution pH value, –O– and –COO– will be converted to –OH or –O–H and generate electrostatic interactions that is favourable to adsorb cationic species. Different categories of constituents wastewater are illustrated in Fig. 1.

4 Water Treatment Technologies: Operational Principles and Limitations

In this section, our consideration will be on two types of water treatment technologies which are; conventional water treatment and advanced water treatment. The conventional water treatment is as follows: physical process which is always referred to as primary or preliminary treatment method, chemical and biological process which are

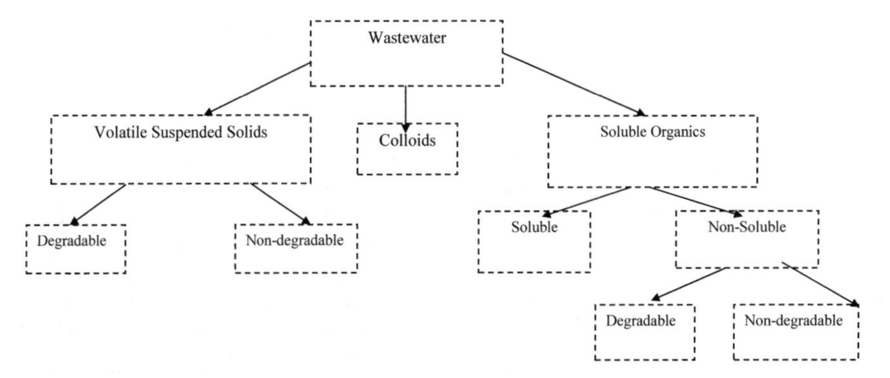

Fig. 1 Organic constituents partition of wastewater

also referred to as secondary treatment methods while the advanced water treatment is refers to as advanced oxidation process called tertiary treatment method. This section gives the reader best understanding of the recent trend on different water treatment technologies. Let's consider these different types of water treatment technologies.

4.1 Conventional Water Treatment Methods

4.1.1 Coagulation and Flocculation

Part of the constituents of most water systems is suspended solids. The purpose of coagulation and flocculation is to separate the suspended solids from the water bodies. Before separation can occur, coagulation reaction must take place by destabilization of colloids through neutralizing their intermolecular force when coagulants are added to the water bodies. There are some factors that are expected to be considered in choosing a coagulant such as particle size, shape, charge, composition and density among others [88]. When particles carry like charges, repulsion occurs and remains suspended in the water. Therefore, the application of the coagulant brings the suspended together to form aggregates in size and settle. The following are several coagulants used for coagulation; iron (II) sulphate, alum and iron (III) Chloride among others [88]. Since the primary aim of coagulation are based on suspended particles and hydrophobic colloids, researchers have tried to improve coagulation for the removal of both insoluble and soluble substances through the grafting of sodium xanthogenate group and it is referred to as amphoteric polyelectrolyte [21]. The pH is one of the main factors to considered through coagulation since when the water pH value is acidic, the negative charges of colloidal substances will be coagulated while there will be removal of cationic ions. When water pH is high, there is a decrease in the turbidity removal which increases the cationic ion's removal. Though coagulation has been helpful in the area of water treatment especially in developing and

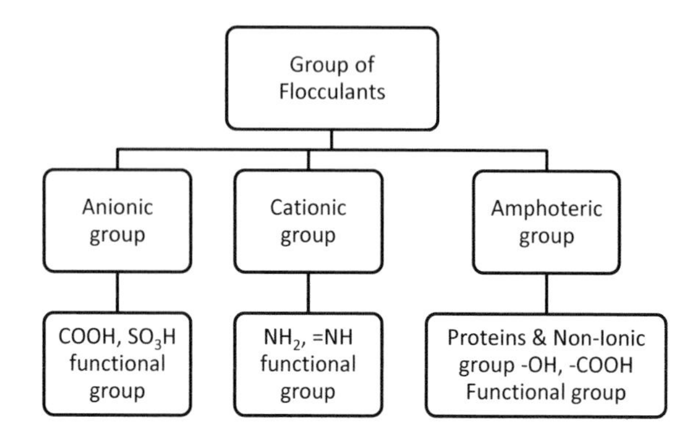

Fig. 2 Groups of Flocculants

under-developing countries, this method is still associated with shortcomings such as formation of sludge which becomes secondary pollutant, high concentration of aluminium residue from the coagulants (i.e. aluminium based) present in the treated water. The presence of aluminium residue is associated with some challenges such as; loss in hydraulic capacity, increased turbidity, reduced disinfection efficiency and potential adverse effects like alzheimer [137]. Though, it can be controlled through adjustment of the pH value to acidic regions. This approach is not generally accepted since it required increase in pH value for the control of corrosion in post treatment in water distribution network which makes the process costly [14].

Flocculation is when polymer act as a bridge between the flocculants and the particles bind into large clumps or agglomerates. Immediately the flocculated suspended particles are formed into larger particles, separation can take place through floatation, straining, or filtration. There are many different types of flocculants namely; polyacrylamide and polyferric sulphate. However, flocculation cannot successfully remove heavy metals directly from wastewater via flocculants. Therefore, researchers have developed macromolecule heavy metals flocculants such as mercaptoacetyl chitosan among others and have been proved to remove heavy metals and also remove turbidity in wastewater (Fig. 2). Apart from heavy metals, differences in the particle's properties are other challenges that make it impossible to use universal flocculant. There are different groups of flocculant; Anionic group contain –COOH, $-SO_3H$ functional group, Cationic groups contain $-NH_2$, =NH functional group, Amphoteric group (both anionic and cationic groups) contain proteins and Nonionic group contain –OH and –COOH functional groups as it is shown in Fig. 2.

Generally, coagulation-flocculation is not enough to treat wastewater completely especially with heavy metals [65]. Therefore, there is a need for other treatment techniques to be applied after coagulation and flocculation in order to achieve complete removal.

4.1.2 Chemical Precipitation

One of the most commonly used wastewater treatment techniques is chemical precipitation which its operation is simple with low cost. This involves the removal of suspended solids, inorganics such as metal, greases, fats, oils and organic substances which will later be separated through filtration or sedimentation. This method is usually used for mitigation of pollutants from the wastewater which contains heavy metals of high concentration but always have low efficiency when there is low concentration [35]. This can result in production of sludge of metallic that need further disposal, making the process of recovery of the metal difficult [60]. There are two types of chemical precipitation namely; Hydroxide and sulphide precipitation.

(I) Hydroxide Precipitation

This is one of the most commonly used chemical precipitation since it is very relative simple, ease pH control and with low cost [36]. The solubility of hydroxides of metals is controlled within the pH values of 8 to 11. Several hydroxides have been applied to precipitate metals from wastewater, since hydroxide precipitation has a low cost and ease of handling, one of the preferred choices of base is lime used in the settings of industrial wastewater [55].

(II) Sulphide Precipitation

This is another form of chemical precipitation which is said to be an efficiency process in removal of toxic heavy metals. Sulphide precipitation is advantageous to hydroxide precipitation because the solubility of its metal sulphide precipitate is lower than the precipitate of hydroxide and its nature is not amphoteric. Therefore, the process of sulphide precipitation was able to achieve more degree of removal of metal ions for a wide range of pH values than hydroxide precipitation. The sludge of metal sulphide also exhibit better characteristics such as dewatering and thickening than the sludge of corresponding hydroxide metals [97]. However, the use of sulphide precipitation comes along with different shortcomings such as production of heavy metal ions and sulphide precipitants in acid conditions whereby sulphide precipitants resulted in evolution of toxic H_2S fumes. It is necessary that sulphide precipitation performed in alkaline or neutral ranges of pH values.

4.1.3 Ultra-Filtration

This is one of membrane filtration techniques used for separation, based on size exclusion mechanism. The separation membrane pore size is within the range of 1–5 μm [148] and falls between nano-filtration and micro-filtration. This method can be used to remove substances of high molecular weight namely; colloidal material, inorganic and organic polymeric molecules. The purpose of these membranes in this technique is to fractionate or separate the wastewater into more useful or less polluting streams but cannot degrade into simpler substances [12]. In view of the above, Ultra-filtration has been increasingly applied in treatment of wastewater. One of the limitations of ultra-filtration is fouling of the membranes which significantly increase trans-membrane pressure and reduced the lifespan of the membrane [7]. There are other several factors that can lead to destruction of the chemical structure of membrane material and membrane fouling such as; concentration polarization, formation of filter cake layer and blockage of membrane pore. Natural organic matter (autochthonous biopolymers and humus) is considered the main culprit for membrane fouling [7].

4.1.4 Reverse Osmosis

The reverse osmosis process is one of the wastewater treatment techniques which involves the use of a pores semi-permeable (that is, cellophane-like) membrane to separate treated water from the wastewater. This technique is capable of removal of dissolved species in a wide range from water which gives more than 20% desalination capacity of the world [19]. There are two sides to this technique namely; treatment side and dilute side. On the treatment side, there is need for the use of pressure on the wastewater forcing the treated water into the dilute side then removed the impurities and washed into the rejected water. Reverse osmosis has been reported for its application in different industries wastewater treatment such as mining, petrochemical, food processing and textile industries [123]. The Membranes with smaller pores size was able to prevent pollutants with larger molecular particle sizes from entering the treated water. However, hazardous chemicals like herbicides, pesticides, insecticides with smaller molecules size than water were able to pass with the treated water [71]. Hence, there is need to incorporate carbon filter with the use of reverse osmosis [71]. There are several challenges associated with reverse osmosis such as been too slow in operation compared with other alternative water treatments, it has high consumption power, its membranes are subjected to decay which leads to poor quality of treated water and also its membrane has shortened life due to treatment of hard water.

4.1.5 Nano-Filtration

This is another membrane filtration technique which is usually an intermediate technique between ultra-filtration and reverse osmosis. Nano-filtration is a promising technology for the treatment of wastewater by removing some pollutants such as dyes, small molecules, nanoparticles and heavy metal ions like copper, chromium, arsenic and nickel among others [115]. There are many benefits associated with Nano-filtration such as reliability, operational ease and comparatively low consumption energy [25]. Nano-filtration application is highly requested for the removal of dyes in wastewater since most dyes average molecular weights is in the range of 100–1000 Da [145]. However, its membranes have pores with low connectivity and low porosity, hence at low pressure the flux is low and the pores block easily [146]. Several researchers had made further studies to improve nano-filtration membranes flux through the properties and structures of the supporting and barrier layers [146].

It has been proved that the application of nano-fibre membrane as a supporting layer should be a good potential candidate for the removal of pollutants.

[33] reported the studies of two commercial nano-filtration namely N30F and NF90 for the removal of pentavalent Arsenic from synthetic water with Arsenic feeds. He further reported that when there is a decrease in the operating temperature with an increase in pH values, the two membranes remove higher concentrations of Arsenic. Feed concentration is one of the parameters which affected Arsenic rejection. [72] also investigated the use of Nano-filtration membrane for the removal of heavy metal ions. He reported that there is rejection of nickel ion through the application of thin-film composite polyamide Nano-filtration membrane. The maximum rejection of nickel was further reported to be 92 and 98% when the initial concentration of the feeds is 250 mg/L and 5 mg/L respectively [72]. He also reported the investigation of binary heavy metals (cadmium and nickel) and the maximum observed solute rejection of cadmium and nickel was 82.69% and 98.94% respectively when the feeds initial concentration is 5 mg/L [72]. There are several reports on heavy metal removal through reverse osmosis membrane and Nano-filtration. [24] used Nano-filtration and reverse osmosis for recovery process copper from wastewater. [59] also studied the efficiency of reverse osmosis membranes and nano-filtration in the removal of metals in the effluent of metallurgical industry. He reported further that the reverse osmosis and nano-filtration desalination satisfied the state reutilization qualification but in industrial large-scale, nano-filtration process would be more suitable.

4.1.6 Solvent Extraction

Another name for solvent extraction is called liquid-liquid extraction. This is a method used in separating different compounds based on their solubility relationship into two different immiscible liquids, usually organic solvent and water. It is also referred to as an extraction process whereby a substance is extracted from one liquid phase to another liquid phase due to differences in its solubility [16]. This can be achieved through application of partition coefficient or poor solubility of two incompatible solvents. It has been widely applied in chemical, food, metallurgical and other industries for its continuous easily operation, perfect process, enhanced economic and superior productivity [142].

Trioctylamine belong to the family of amine and its molecular formula is $(CH_3-(CH_2)_7)_3N$ and possess high basicity [142]. Trioctylamine has the ability to generate complex anions and to gives back-extraction through its combination with metal ions which can easily recover organic acids from aqueous solution [142]. It has many benefits such as high extraction capacity, good extraction kinetics and negligible solubility in the aqueous phase among others [142]. Thus, it was commonly applied in the recovery process of much precious metal and organic acid.

4.1.7 Electrolysis

Electrolysis is one of the treatment techniques used in metals removal in wastewater. This process involves the generation of current from electricity and passes via aqueous medium which contains an insoluble anode and cathode plate. This electricity process can be produced through movement of electrons from an element to another. When electrochemical process is used for heavy metal removal in wastewater, it precipitates in neutralized or weak acidic hydroxides. Electrochemical treatments in wastewater includes; electro-coagulation, electro-deposition, electro-oxidation and electro-flotation [40]. Electro-coagulation process is formed through electrolytic oxidation of an appropriate material of anode. In this process, the removal of charged metal species from the wastewater is allowed by reacting with the effluent anion. The advantages of this process are; no required use of chemicals, reduction in the production of sludge and ease of operation [40].

4.1.8 Adsorption Technology

This is a technique that involves accumulation of solutes that is liquid or solid on the site surface of an adsorbent (liquid or solid) to generate an atomic or molecular film [76]. The nature of adsorption operation is more of a biological or chemical or physical system that is commonly applied industrially for water treatment and activated charcoal among others. Due to its low cost and simplicity, adsorption has been considered suitable techniques required in treatment of wastewater [4]. This technology has been widely used for the removal of heavy metal ions from different industrial wastewaters [1]. One of the most commonly applied adsorbents is Activated carbon. This adsorbent has been confirmed by many authors dues to its highly porous and amorphous crystalline solid nature. Other commonly used adsorbents are clay minerals [77] zeolites, industrial solid wastes and biomaterials [140]. Certain waste from the operation of agriculture or natural materials is one of the low cost adsorbents used. Generally, most of these materials are locally available in large quantities as waste and contributed to pollution of the ecosystem. Therefore, they are very cheap. However, regeneration of the adsorbents after usage remains a challenge facing adsorption technology.

4.2 Advanced Oxidation Processes

4.2.1 Photocatalysis

Photocatalysis process is a chemical reaction that involves radiation of catalyst material using photon [85]. When the catalyst and the system are in the same phase, it is referred to as homogeneous photocatalysis while when the catalyst and the system are in different phases, it is referred to as heterogeneous photocatalysis [11]. Photocatalytic technique is the formation of hole-electron pairs through the absorption of light of photocatalyst; the radicals include hydroxyl, hydroperoxyl and superoxide are been formed through the photogenerated charges [119]. These radicals react via reduction and oxidation reactions, reducing the organic compound and inorganic ions and oxidizing the organic compounds [75]. There are several other ways whereby photocatalysis is applied such as; environmental surfaces and medical devices [102], environmental cleaning [18], coating for hard surfaces in hospital environment [23], self-disinfecting catheters [93], percutaneous implant [107], metal implant [22], intraocular lenses [44], dental implant [26], dental adhesives [106], antimicrobial healthcare surfaces [2] among others.

In the process of heterogeneous photocatalysis, the semiconductor employed as a photocatalyst is very important. The most suitable photocatalyst are metals of oxide or metals of sulphide which have band gap energy suitable to transit electrons by the source of irradiation [84]. Some of these semiconductors include TiO_2, ZnO, WO_3, CdS, SiO_2 and Fe_2O_3 among others. There is formation of a positive empty site when an electron leaves the valence band to conduction band. This empty site is referred to as hole (h^+). The recombination of the electron-hole pair photogenerated occurs either through material surface or bulk, whereby moves to a state of initiation and dissipate the heat (energy) [91]. Also, it can further react with the compounds adsorbed on the surface of the photocatalyst by donating or accepting electrons or trapped in the metastable surface state. It is advantageous that the material for photocatalyst can be excited through cheaply available sources like sun radiation. Several researchers have reported mineralization of the toxic organic compound present in the wastewater into water, carbon dioxide and simple inorganic acids [111]. Therefore, photocatalytic method can break down dangerous and toxic compounds like phenol, biphenyl, pesticides, solvents, dioxins and dyes among others to generate non-toxic mineral species namely carbon (IV) oxide and water.

$$\text{Organicpollutant} + O_2 \overset{\text{Photocatalyst/Hv}}{\rightarrow} \text{CO}_2 + \text{H}_2\text{O} + \text{mineralacids}$$

Reaction mechanism of photocatalysis process

Photocatalysis principle involves photocatalyst excitation when exposed to light radiation namely Ultraviolet light or visible light. During photon's action, the photocatalyst generate highly oxidizing free radicals which further involves breaking of the chemicals adsorbed on its surface [70]. The photocatalyst converts from chemical energy to photon energy via reduction-oxidation reaction which results in the nanoparticle activation and degradation of the present molecular chemical. This degradation process involves the succession of radical oxidation initiated via oxidants like OH^o which is directly produced via the light degradation of water molecules adsorbed on the photocatalyst active site. The followings are some of the parameters considered in the photolysis of organic compounds by a semiconductor;

i. The water content and the nature of the formation of hydroxyl radicals and concentration of pollutants.
ii. The intensity and nature of irradiation of the source of light.
iii. The numerical value of incident rays to the nanoparticle activation.
iv. The reaction medium is determined by the amount of the actives site on the nanoparticle.

The processes can be summarized into four levels:

A. Nanoparticle (NP) activation

This level is associated with activation process through the illuminating of the nanoparticle which generates the hole-electron pair.

$$NP + hv \rightarrow NP + e^-_{BC} + h^+_{BV}$$

B. Electron and Holes separation

The hole/electron pairs duration (H^+/e^-) is very short and their recombination is through heat. When the electron donor and acceptor are not present, the recombination process is very fast.

$$NP\left(e^-_{BC} + h^+_{BV}\right) \rightarrow \text{liberationofEnergy}$$

C. Oxidation and Reduction

This reaction takes place at the semiconductor's surface when there is a reaction between the positive holes and the water molecule which produces a radical of hydroxyl. These reactions of oxidation or reduction are what brought about treatment of wastewater.

Reduction Reaction

$$O_{2,ads} + e^-_{BC} \rightarrow O^-_{2,ads}$$

Oxidation reaction

$$H_2O + h^+_{BV} \rightarrow OH^{\circ}_{ads} + H^+$$

$$OH^-_{ads} + h^+_{BV} \rightarrow OH^{\circ}_{ads}$$

$$R_{ads} + h^+_{BV} \rightarrow R^{\circ}_{ads}$$

Band Gap Photo-excitation

This is referred to as the energy changes which result from the difference between conduction band (Bottom) and valence band (top) of a semiconductor or photocatalyst. Photocatalysis process is referred to as photocatalyst excitation which occurs through the illumination of light energy which may be equal to or greater than their band gap. Semiconductors with their band gaps are listed as follows: TiO_2 (3.43 eV) [89], ZnO (3.37 eV) [114], WO_3 (2.8 eV) [144], CdS (2.42 eV) [73], SnO_2 (3.6 eV) [57] and GaN (3.39 eV) [96] among others. It has been observed that the gap of the band is always depending on shape and the size of particles of the nanoparticles. From several studies, when the energy of the band gap increases, the semiconductor particle size also decreases. Among several reported semiconductors, TiO_2 has been treated promising and proven as the most suitable semiconductor for its application in ecosystem due to its non-toxicity, chemical stability, physical and convenience for use at ambient temperature and pressure [38]. Anatase TiO_2 with 3.2 eV has been activated under ultraviolet irradiation which takes about 5% of solar energy. Researchers have made some efforts to develop photocatalyst which required low energy with visible light via engineering of band doping [30]. Doping of transition metal such as cations namely; V, Fe, Cr and Ni on the site of Ti [141] and anions namely; S, N and C on the site of O [66] which are usually synthesized to improve the activity of photocatalytic reaction. [30] reported that synthesized of N + Zr co-doped TiO_2 exhibited excellent photocatalytic efficiency than Zr or N-doped. Non-metal ions N, C on the site of TiO_2 have been used to reduce the gap of band of the TiO_2.

However, there are different challenges attached with mono-doping of TiO_2 induced into the system such as; unstability, reducing the current of the photo-excited or carrier mobility, recombination of photo-excited electron-hole pairs [54] which is unfavourable to improve the efficiency of photocatalytic reaction.

4.2.2 Photo-Fenton

Photo-Fenton reaction is one of the commonly studied advanced oxidation processes used for the removal of emerging pollutants in wastewater [80]. Photo-Fenton reaction leads to generation of hydroxyl radical from hydrogen peroxide. Several studies have found this technique effective for the removal of many classes of pollutants, such as pesticides [28], dyes [110], insecticides [29], pharmaceuticals [34], chlorophenol [112], Nitrobenzene [64] and polychlorinated biphenyls [42]. Furthermore, economic and environmental sustainability have been improved through the employment of solar energy in this process. However, this application has some setbacks which include a strict pH control to ensure that the catalyst (Fe(II) and Fe(III) species) plays their roles, by preventing formation of inactive iron oxyhydroxide precipitation so as to have optimum concentration of photoactive species. Therefore, there is a need to operate in a narrow pH range of 2.8–3.5.

Photo-Fenton reaction mechanism

Photo-Fenton reaction can be identified by reactions 1 and 2 below;

$$Fe^{3+} + H_2O + hv \rightarrow Fe^{2+} + HO + H^+$$

$$H_2O_2 + Fe^{2+} \rightarrow Fe^{3+} + HO + OH^-$$

This reaction involves several stages based on the role in the reactions of these free radicals: (I) generation of active oxygen with the presence of oxidation species (that is, species that initiate oxidation reaction); (II) species transformation are present in active oxygen; (III) transformation reaction of species of oxygen with organic compound; (IV) intermediate reaction termination. The ferrous ions generated through photolysis entered Fenton reaction and generated supplemental radicals called hydroxyl. Consequently, the rate of oxidation of photo-Fenton is accelerated compared to Fenton process [41]. In addition, when there is high rate of oxidation, the total iron utilization and the generation of sludge are always reduced in photo-Fenton when compared with Fenton process [133]. Furthermore, it has been established that photo-Fenton via UV or solar light, has greater effects on inactivation of microorganisms in wastewater. However, the important of the technique is determined by microorganism present [87] and the nature of water under treatment [121]. In a study, it was found that the spores of *F. solani* is more resistant to the treatment by solar than the *E. coli* vegetative cells [39]. High oxidation like Hydroxyl radicals with a potential of 2.80 V, was said to degrade compounds of recalcitrant like chlorinated and phenolic compounds [47].

4.2.3 Photoelectrocatalysis

Generally, photoelectrocatalysis systems consist of photocatalysis as an anode material and exposed to light source (UV light or sun light) for effective breaking down of organic pollutants in the aqueous medium. Photoelectrocatalysis is more advantageous than photocatalysis due to easy separation of holes and electrons in the presence of electrical charges [83]. Graphene oxide based nanocomposites have been employed in Photoelectrocatalysis as working electrode and its performance has improved significantly [139] synthesized ternary polyaniline—graphene oxide—titanium oxide hybrid films and applied as photoanode in the photoelectrocatalytic degradation for reduction of water. When compare with pristine titanium oxide electrodes, the hybrid films with alternative graphene oxide, titanium oxide and polyaniline layers exhibited better photoelectrocatalysis activity with higher rate of hydrogen compared to most reported titanium oxide based nanophotoctalyst. The mechanism of Photoelectrocatalysis suggested that the efficient charged transfer occurred at the interface of polyaniline and graphene oxide—titanium oxide due to the potential of the matched band [139]. Polyaniline served as both electron transporter and collector and the transfer of the collected electrons to counter electrode was to form H_2 through the reduction of water under a bias voltage [139].

Photoelectrocatalysis mechanism

This technique involves combination of both photocatalytic and electrolytic processes for recombination of electron-hole (e^-_{CB}/h^+_{VB}) pairs and increasing the lifespan of the holes [99]. The fundamental procedures of photolelectrocatalysis are ejection of an electron (e_{CB}^{-}) through the valence band of a semiconductor which completely occupied conduction band and formed a positively charged vacancy (h_{VB}^{+}). When the semiconductor is being exposed to illumination energy which is greater than its band gap, it gives rise to the photo-excitation of the e_{CB}^{-} to conductive band from valence band. Then, the light allows formation of (e_{CB}^{-}/h_{VB}^{+}) pairs through Reaction (1) [78],

$$\mathrm{NP} + h_{v} \rightarrow e_{CB}^{-} + h_{VB}^{+}$$

Considering reaction 1, in the product, photogenerated h_{VB}^{+} is a strong oxidizing species and e_{CB}^{-} is the potential reductor. The photogenerated h_{VB}^{+} then oxidized the organic pollutants till their mineralization completed. In reaction 2, h_{VB}^{+} adsorbed water to generate strong oxidant °OH during mineralizes of the organic pollutants. Then in reaction 3, e_{CB}^{-} react with the adsorbed O_2 to produce the superoxide radical $O_2^{\cdot-}$.

$$h_{VB}^{+} + \mathrm{H_2O} \rightarrow^{\circ} \mathrm{OH} + \mathrm{H}^{+}$$

$$e_{CB}^{-} + \mathrm{O_2} \rightarrow \mathrm{O_2^{\cdot-}}$$

Other reactive oxygen species are formed through reactions 4 and 5. The following reactive oxygen species are; H_2O_2, OH^{-} and $HO_2^{\cdot}$.

$$\mathrm{O_2^{\cdot-}} + \mathrm{H}^{+} \rightarrow \mathrm{HO_2^{\cdot}}$$

$$\mathrm{HO_2^{\cdot}} \rightarrow \mathrm{H_2O_2} + \mathrm{O_2}$$

In reaction 6, e_{CB}^{-} is an unstable species at the excited state which tends to return to the group state either with adsorbed °OH or via recombination with the unreacted h_{VB}^{+}

$$e_{CB}^{-} +^{\circ} \mathrm{HO} \rightarrow \mathrm{OH}^{-}$$

$$e_{CB}^{-} + h_{VB}^{+} \rightarrow \mathrm{NP} + \mathrm{Heat}$$

4.2.4 Catalytic Ozonation

Ozonation is considered one of the most commonly used wastewater treatment processes. This process was been employed to oxidized pollutants with a large spectrum of emerging concerns and to dissolved organic matters. Ozone is the main agent applied in ozonation. It is selective whereby react with electron rich moieties.

Ozonation partly oxidize chemical compounds and emits toxic by-products, such as bromate from bromine containing wastewater, which is carcinogenic [48]. Due to the need to control the release of toxic emission and reactive by-products, post treatment is required whereby activated carbon or sand filtration are applied after ozonation [113]. But [52] noted that some ozonation by-products are incompletely removed even during biological post-treatment steps.

Recently, researchers found that catalytic ozonation (combination of ozonation and catalysis) have overcome several disadvantages of ozonation process such as energy requirement and higher operational cost. Heidarizad [50] reported catalytic ozonation has a promising advanced oxidation process for degradation of pollutants in wastewater. Heidari [49] found the modified Graphene oxide with magnesium oxide also give high adsorption capacity of Methylene blue. The data reviewed that there was significant mineralization compare to that of ozonation which is without catalyst.

Heterogeneous and homogeneous catalytic ozonation

The fundamental aspect of the mechanism for homogenous catalytic ozonation is on decomposition reaction of ozone which produced radicals called hydroxyl. The metal ions speed up the decomposition process to generate $\cdot O^{2-}$ which further transfer to O_3 to gain $\cdot O_3$. Several researchers have proposed schemes for the mechanism of homogenous catalytic ozonation. One of them is [86] which proposed that a system of cobalt (II) oxalate/ozone at acidic pH (6) [130]. He proposed the first stage in the reaction pathway to form cobalt (II) oxalate and the centre of the metal was established to be site of attack. The electron density is partially donated to Cobalt (II) from oxalate which further caused the cobalt (II) oxalate reactivity to increase when compared with the free cobalt (II). This process can effectively improve the organic pollutants removal efficiency but its shortcoming is the introduction of ions which leads to secondary pollution and further increase the water treatment costs.

While for heterogeneous catalytic ozonation; there are three different phases in consideration such as; liquid, gaseous and solid which involves aqueous catalytic ozonation due to it heterogeneous nature of the process. Catalytic ozonation techniques are also affected by some shortcomings such as; target pollutants, nature of catalysts and the range of pH of the solution. Hence, it is not easy to develop efficient techniques to determine the reactive intermediate process. There are two typical mechanisms are as following.

i. Interfacial reaction mechanism

The catalysts first adsorb by its surface to mitigate organic pollutants. Then generate active site to react with the target pollutants to produced active complexes which make the pollutants oxidized via ozone or HO^{o} present in the solution and the catalyst surface oxidize the intermediate or desorbed in order to oxidized via ozone or HO^{o} [13].

Legube and Leitner [62] reported the use of Me–OH catalyst as an adsorbent only; where ozone and hydroxyl radical are oxidant species. The adsorption process of organic acid first takes place on the catalyst. Then, strong anions appear and formed five or six membered chelate rings on the surface. While the hydroxyl or ozone radical oxidized complex surface and produce a by-product of oxidation. In this mechanism, the catalysts act as an adsorptive material that combines with the organic pollutants to generate chelate which can break down through HO° or ozone radical. However, there are some heterogeneous catalytic processes that have poor adsorptive ability and are difficult to explain via this mechanism.

ii. HO° Mechanism

This mechanism proposed that the solubility of ozone and initiates its decomposition can be increased through metal oxide catalyst. The hydroxyl functional group presents on the surface of the metal oxide play a crucial role in the formation of HO°. The surface of the catalyst adsorbed the soluble ozone in aqueous solution, transfer of series of radical chains to produce additional HO° which give high potential oxidation and can oxidize the pollutants in the wastewater. Poor adsorptive ability of catalytic ozonation system can be explained via this mechanism. Legube and Leitner [62], reported that in this mechanism, the catalyst reacts with the adsorbed organic pollutants and ozone together. Hence, the catalyst is being reduced as the ozone oxidized the metal. The reaction of ozone on the reduced metal leads to HO radicals. The organic acid was adsorbed on the oxidized catalyst and then oxidized by an electron—transfer reaction to gain again reduced catalyst.

5 Conclusion

In this review, physico-chemical parameters for wastewater and classification of pollutants in wastewater were first examined, certain aspects such as physical, chemical characteristics, organic and inorganic pollutants were discussed. The potentials of conventional water treatment methods to purify wastewater successfully along with their shortcoming were discussed. When comparing conventional water treatment with advanced water treatment methods, it was established that the advanced water treatment method performed better and successfully more than conventional water treatment especially when there are multi-pollutants in wastewater. Research trends and knowledge gaps are in the area of hybrid advanced water treatment where multi-advanced will be applied.

References

1. Abdullahi A, Ighalo J, Ajala O (2020) Physicochemical analysis and heavy metals remediation of pharmaceutical industry effluent using bentonite clay modified by H_2SO_4 and HCl. JOTCSA 7(3):17. https://doi.org/10.18596/jotcsa.703913
2. Adlhart C, Verran J, Azevedo NF, Olmez H, Keinänen-Toivola MM, Gouveia I, Crijns F (2018) Surface modifications for antimicrobial effects in the healthcare setting: a critical overview. J Hosp Infect 99(3):239–249

3. Ajala OJ, Ighalo JO, Adeniyi AG, Ogunniyi S, Adeyanju CA (2020) Contamination issues in sachet and bottled water in Nigeria: a mini-review. Sustain Water Resour Manage 6(6):1–10
4. Ajala OJ, Nwosu FO, Ahmed RK (2018) Adsorption of atrazine from aqueous solution using unmodified and modified bentonite clays. Appl Water Sci 8(7):214
5. Ajayi A, Peter-Albert C, Ajojesu T, Bishop S, Olasehinde G, Siyanbola T (2016) Biochemical oxygen demand and carbonaceous oxygen demand of the covenant university sewage oxidation pond. Covenant J Phys Life Sci 4(1):11–19
6. Ajibade FO, Adelodun B, Lasisi KH, Fadare OO, Ajibade TF, Nwogwu NA, Wang A (2020) Environmental pollution and their socioeconomic impacts. Microbe mediated remediation of environmental contaminants. Elsevier, pp 321–354
7. Akhondi E, Zamani F, Law AW, Krantz WB, Fane AG, Chew JW (2017) Influence of backwashing on the pore size of hollow fiber ultrafiltration membranes. J Membr Sci 521:33–42
8. Alharbi OM, Khattab RA, Ali I (2018) Health and environmental effects of persistent organic pollutants. J Mol Liq 263:442–453
9. Ansari FA, Gupta SK, Nasr M, Rawat I, Bux F (2018) Evaluation of various cell drying and disruption techniques for sustainable metabolite extractions from microalgae grown in wastewater: a multivariate approach. J Clean Prod 182:634–643
10. Atay E, Ertekin A, Bozkurt E, Aslan E (2020) Impact of bisphenol A on neural tube development in 48-hr chicken embryos. Birth Defects Res 112(17):1386–1396
11. Augugliaro V, Palmisano G, Palmisano L, Soria J (2019) Heterogeneous photocatalysis and catalysis: an overview of their distinctive features heterogeneous photocatalysis. Elsevier, pp 1–24
12. Awaleh MO, Soubaneh YD (2014) Waste water treatment in chemical industries: the concept and current technologies. Hydrol Current Res 5(1). https://doi.org/10.4172/2157-7587.1000164
13. Bao Q (2019) Catalytic ozonation of aromatics in aqueous solutions over graphene and their derivatives. A new generation material graphene: applications in water technology. Springer, pp 209–219
14. Birnhack L, Penn R, Lahav O (2008) Quality criteria for desalinated water and introduction of a novel, cost effective and advantageous post treatment process. Desalination 221(1–3):70–83
15. Bolzonella D, Papa M, Da Ros C, Anga Muthukumar L, Rosso D (2019) Winery wastewater treatment: a critical overview of advanced biological processes. Crit Rev Biotechnol 39(4):489–507
16. Boo C, Billinge IH, Chen X, Shah KM, Yip NY (2020) Zero liquid discharge of ultrahigh-salinity brines with temperature swing solvent extraction. Environ Sci Technol 54(14):9124–9131
17. Cai H, Liang J, Ning X-A, Lai X, Li Y (2020) Algal toxicity induced by effluents from textile-dyeing wastewater treatment plants. J Environ Sci 91:199–208
18. Cai T, Liu Y, Wang L, Zhang S, Zeng Y, Yuan J, Luo S (2017) Silver phosphate-based Z-Scheme photocatalytic system with superior sunlight photocatalytic activities and anti-photocorrosion performance. Appl Catal B 208:1–13
19. Caldera U, Breyer C (2017) Learning curve for seawater reverse osmosis desalination plants: capital cost trend of the past, present, and future. Water Resour Res 53(12):10523–10538
20. Capodaglio AG (2020) Fit-for-purpose urban wastewater reuse: Analysis of issues and available technologies for sustainable multiple barrier approaches. Crit Rev Environ Sci Technol 1–48
21. Chang Q, Wang G (2007) Study on the macromolecular coagulant PEX which traps heavy metals. Chem Eng Sci 62(17):4636–4643
22. Chen S, Zhou Y, Li J, Hu Z, Dong F, Hu Y, Wu Z (2020) Single-atom Ru-implanted metal-organic framework/MnO_2 for the highly selective oxidation of NOx by plasma activation. ACS Catalysis 10(17):10185–10196
23. Clemente A, Ramsden JJ, Wright A, Iza F, Morrissey JA, Puma GL, Malik DJ (2019) Staphylococcus aureus resists UVA at low irradiance but succumbs in the presence of TiO_2 photocatalytic coatings. J Photochem Photobiol, B 193:131–139

24. Cséfalvay E, Pauer V, Mizsey P (2009) Recovery of copper from process waters by nanofiltration and reverse osmosis. Desalination 240(1–3):132–142
25. Darban A, Shahedi A, Taghipour F, Jamshidi-Zanjani A (2020) A review on industrial wastewater treatment via electrocoagulation processes. Curr Opin Electrochem
26. Dini C, Nagay BE, Cordeiro JM, da Cruz NC, Rangel EC, Ricomini-Filho AP, Barão VA (2020) UV-photofunctionalization of a biomimetic coating for dental implants application. Mater Sci Eng C 110:110657
27. Dube K, Nhamo G (2019) Evidence and impact of climate change on South African national parks. Potential implications for tourism in the Kruger National Park. Environ Develop 100485
28. Dutta A (2017) Solar photo fenton degradation of AZO dye and pesticide in wastewater study of batch and continuous processes
29. Dutta A, Datta S, Ghosh M, Sarkar D, Chakrabarti S (2017) Sunlight-assisted photo-fenton process for removal of insecticide from agricultural wastewater. Trends in Asian water environmental science and technology. Springer, pp 23–33
30. Eldeeb MS, Fadlallah MM, Martyna GJ, Maarouf AA (2018) Doping of large-pore crown graphene nanomesh. Carbon 133:369–378
31. Ertani A, Mietto A, Borin M, Nardi S (2017) Chromium in agricultural soils and crops: a review. Water Air Soil Pollut 228(5):190
32. Figinsky FR (2016) Odor Monitoring at the New Orleans east bank wastewater treatment plant
33. Figoli A, Cassano A, Criscuoli A, Mozumder MSI, Uddin MT, Islam MA, Drioli E (2010) Influence of operating parameters on the arsenic removal by nanofiltration. Water Res 44(1):97–104
34. Foteinis S, Monteagudo JM, Durán A, Chatzisymeon E (2018) Environmental sustainability of the solar photo-Fenton process for wastewater treatment and pharmaceuticals mineralization at semi-industrial scale. Sci Total Environ 612:605–612
35. Fu F, Wang Q (2011) Removal of heavy metal ions from wastewaters: a review. J Environ Manage 92:407–418. https://doi.org/10.1016/j.jenvman.2010.11.011
36. Fu F, Xie L, Tang B, Wang Q, Jiang S (2012) Application of a novel strategy—advanced Fenton-chemical precipitation to the treatment of strong stability chelated heavy metal containing wastewater. Chem Eng J 189:283–287
37. Fukushima T, Inomata T, Komatsu E, Matsushita B (2019) Factors explaining the yearly changes in minimum bottom dissolved oxygen concentrations in Lake Biwa, a warm monomictic lake. Sci Rep 9(1):1–10
38. Gao C, Wei T, Zhang Y, Song X, Huan Y, Liu H, Chen X (2019) A photoresponsive rutile TiO_2 heterojunction with enhanced electron-hole separation for high-performance hydrogen evolution. Adv Mater 31(8):1806596
39. Giannakis S, López MIP, Spuhler D, Pérez JAS, Ibáñez PF, Pulgarin C (2016) Solar disinfection is an augmentable, in situ-generated photo-Fenton reaction—part 2: a review of the applications for drinking water and wastewater disinfection. Appl Catal B 198:431–446
40. Gunatilake S (2015) Methods of removing heavy metals from industrial wastewater. Methods 1(1):14
41. Guo Q, Li G, Liu D, Wei Y (2019) Synthesis of zeolite Y promoted by Fenton's reagent and its application in photo-Fenton-like oxidation of phenol. Solid State Sci 91:89–95
42. Gutiérrez-Hernández RF, Bello-Mendoza R, Hernández-Ramírez A, Malo EA, Nájera-Aguilar HA (2019) Photo-assisted electrochemical degradation of polychlorinated biphenyls with boron-doped diamond electrodes. Environ Technol 40(1):1–10
43. Halim A, Sharmin S, Rahman H, Haque M, Rahman S, Islam S (2018) Assessment of water quality parameters in baor environment, Bangladesh: a review. Int J Fisher Aquatic Stud 6(2):269–263
44. Hampp N, Dams C, Badur T, Reinhardt H (2017) TiO_2 nanoparticles for enhancing the refractive index of hydrogels for ophthalmological applications. Paper presented at the Colloidal Nanoparticles for Biomedical Applications XII.

45. Han F, Zong Y, Jassby D, Wang J, Tian J (2020) The interactions and adsorption mechanisms of ternary heavy metals on boron nitride. Environ Res 183:109240
46. Han L, Li B, Liu R, Peng J, Song Y, Wang S, Zhang M (2018) Management technology and strategy for environmental risk sources and persistent organic pollutants (POPs) in Liaohe River Basin. Chin Water Syst. Springer, pp 273–347
47. Hassaan MA, El Nemr A (2017) Advanced oxidation processes for textile wastewater treatment. Int J Photochem Photobiol 2(3):85–93
48. Heeb MB, Criquet J, Zimmermann-Steffens SG, von Gunten U (2014) Oxidative treatment of bromide-containing waters: Formation of bromine and its reactions with inorganic and organic compounds. Water Resour 48:15–42
49. Heidari MR, Varma RS, Ahmadian M, Pourkhosravani M, Asadzadeh SN, Karimi P, Khatami M (2019) Photo-fenton like catalyst system: activated carbon/$CoFe_2O_4$ nanocomposite for reactive dye removal from textile wastewater. Appl Sci 9(5):963
50. Heidarizad M, Şengör SS (2017) Graphene oxide/magnesium oxide nanocomposite: a novel catalyst for ozonation of phenol from wastewater. Paper presented at the World Environmental and Water Resources Congress 2017
51. Heldele A-S (2017) Solubility of arsenic in Swedish contaminated soils
52. Hübner U, von Gunten U, Jekel M (2015) Evaluation of the persistence of transformation products from ozonation of trace organic compounds: a critical review. Water Resour 68:150–170
53. Ijoma G, Tekere M (2017) Potential microbial applications of co-cultures involving ligninolytic fungi in the bioremediation of recalcitrant xenobiotic compounds. Int J Environ Sci Technol 14(8):1787–1806
54. Islam MN (2019) The role of aluminum, cobalt co-doping on the band gap tuning of TiO_2 thin film deposited by spray pyrolysis. Bangladesh University of Engineering and Technology
55. Kavak D (2013) Removal of lead from aqueous solutions by precipitation: statistical analysis and modeling. Desalination Water Treat 51(7–9):1720–1726
56. Kenawy I, Hafez M, Ismail M, Hashem M (2018) Adsorption of Cu(II), Cd(II), Hg(II), Pb(II) and Zn(II) from aqueous single metal solutions by guanyl-modified cellulose. Int J Biol Macromol 107:1538–1549
57. Kim D, Kim DH, Riu D-H, Choi BJ (2018) Temperature effect on the growth rate and physical characteristics of SnO_2 thin films grown by atomic layer deposition. Arch Metallurgy Mater 63
58. Koegel-Knabner I, Rumpel C (2018) Advances in molecular approaches for understanding soil organic matter composition, origin, and turnover: a historical overview. Advances in agronomy, vol 149. Elsevier, pp 1–48
59. Koseoglu H, Kitis M (2009) The recovery of silver from mining wastewaters using hybrid cyanidation and high-pressure membrane process. Miner Eng 22(5):440–444
60. Lakherwal D (2014) Adsorption of heavy metals: a review. Int J Environ Res Develop 4(1):41–48
61. Lal S, Singhal A, Kumari P (2020) Exploring carbonaceous nanomaterials for arsenic and chromium removal from wastewater. J Water Process Eng 36:101276
62. Legube B, Leitner NKV (1999) Catalytic ozonation: a promising advanced oxidation technology for water treatment. Catal Today 53(1):61–72
63. Lellis B, Fávaro-Polonio CZ, Pamphile JA, Polonio JC (2019) Effects of textile dyes on health and the environment and bioremediation potential of living organisms. Biotechnol Res Innov 3(2):275–290
64. Liu F, Yao H, Sun S, Tao W, Wei T, Sun P (2020) Photo-fenton activation mechanism and antifouling performance of an FeOCl-coated ceramic membrane. Chem Eng J 125477
65. López-Maldonado E, Oropeza-Guzman M, Jurado-Baizaval J, Ochoa-Terán A (2014) Coagulation–flocculation mechanisms in wastewater treatment plants through zeta potential measurements. J Hazard Mater 279:1–10
66. Luitel H, Chettri P, Tiwari A, Sanyal D (2019) Experimental and first principle study of room temperature ferromagnetism in carbon-doped rutile TiO_2. Mater Res Bull 110:13–17

67. Lv W, Yuan Q, Lv W, Zhou W (2020) Effects of introducing eels on the yields and availability of fertilizer nitrogen in an integrated rice–crayfish system. Sci Rep 10(1):1–8
68. Ma J, Wu S, Shekhar N, Biswas S, Sahu AK (2020) Determination of physicochemical parameters and levels of heavy metals in food waste water with environmental effects. Bioinorganic Chem Appl
69. Madikizela LM, Ncube S, Tutu H, Richards H, Newman B, Ndungu K, Chimuka L (2020) Pharmaceuticals and their metabolites in the marine environment: sources, analytical methods and occurrence. Trends Environ Anal Chem e00104
70. Mamaghani AH, Haghighat F, Lee C-S (2017) Photocatalytic oxidation technology for indoor environment air purification: the state-of-the-art. Appl Catal B 203:247–269
71. Morin OJ (2011) Principles and practises of reverse osmosis-encyclopedia of life support systems. Membrane Process 11
72. Murthy Z, Chaudhari LB (2008) Application of nanofiltration for the rejection of nickel ions from aqueous solutions and estimation of membrane transport parameters. J Hazard Mater 160(1):70–77
73. Muruganandam S, Anbalagan G, Murugadoss G (2017) Structural, electrochemical and magnetic properties of codoped (Cu, Mn) CdS nanoparticles with surfactant PVP. Optik 131:826–837
74. Nadarajan S, Sukumaran S (2020) Chemistry and toxicology behind chemical fertilizers. Controlled release fertilizers for sustainable agriculture. Elsevier
75. Ng NL, Brown SS, Archibald AT, Atlas E, Cohen RC, Crowley JN, Fuchs H (2017) Nitrate radicals and biogenic volatile organic compounds: oxidation, mechanisms, and organic aerosol. Atmos Chem Phys 17(3):2103
76. Nwosu FO, Ajala OJ, Okeola FO, Adebayo SA, Olanlokun OK, Eletta AO (2019) Adsorption of chlorotriazine herbicide onto unmodified and modified kaolinite: Equilibrium, kinetic and thermodynamic studies. Egypt J Aquatic Res 45(2):99–107
77. Nwosu FO, Ajala OJ, Owoyemi RM, Raheem BG (2018) Preparation and characterization of adsorbents derived from bentonite and kaolin clays. Appl Water Sci 8(7):195
78. Ojha N, Bajpai A, Kumar S (2019) Visible light-driven enhanced CO_2 reduction by water over Cu modified S-doped gC_3N_4. Catal Sci Technol 9(17):4598–4613
79. Okon B, Okon V, Udom E (2020) Water infrastructure maintainability–issues and challenges in the coastal regions of Akwa Ibom State. Nigerian J Technol 39(3):953–961
80. Ortiz I, Rivero MJ, Margallo M (2019) Advanced oxidative and catalytic processes. Sustainable water and wastewater processing. Elsevier, pp 161–201
81. Otero XL, De La Peña-Lastra S, Pérez-Alberti A, Ferreira TO, Huerta-Diaz MA (2018) Seabird colonies as important global drivers in the nitrogen and phosphorus cycles. Nat Commun 9(1):1–8
82. Padhan S, Karthik R, Padmavati G (2019) Diurnal variation of phytoplankton community in the coastal waters of South Andaman Island with special emphasis on bloom forming species
83. Pan L, Sun S, Chen Y, Wang P, Wang J, Zhang X, Wang ZL (2020) Advances in piezo-phototronic effect enhanced photocatalysis and photoelectrocatalysis. Adv Energy Mater 10(15):2000214
84. Parrino F, Loddo V, Augugliaro V, Camera-Roda G, Palmisano G, Palmisano L, Yurdakal S (2019) Heterogeneous photocatalysis: guidelines on experimental setup, catalyst characterization, interpretation, and assessment of reactivity. Cataly Rev 61(2):163–213
85. Pathak N, Caleb OJ, Geyer M, Herppich WB, Rauh C, Mahajan PV (2017) Photocatalytic and photochemical oxidation of ethylene: potential for storage of fresh produce—a review. Food Bioprocess Technol 10(6):982–1001
86. Pines DS, Reckhow DA (2002) Effect of dissolved cobalt (II) on the ozonation of oxalic acid. Environ Sci Technol 36(19):4046–4051
87. Polo-López MI, Pérez JAS (2020) Perspectives of the solar photo-Fenton process against the spreading of pathogens, antibiotic resistant bacteria and genes in the environment. Curr Opin Green Sustain Chem 100416

88. Prakash NB, Sockan V, Jayakaran P (2014) Waste treatment by coagulation and flocculation. Int J Eng Sci Innov Technol 3(2):479
89. Prasad S, Kumar SS, Shajudheen VM (2020) Synthesis, characterization and study of photocatalytic activity of TiO_2 nanoparticles. Mater Today Proc
90. Prüss-Ustün A, Wolf J, Bartram J, Clasen T, Cumming O, Freeman MC, Johnston R (2019) Burden of disease from inadequate water, sanitation and hygiene for selected adverse health outcomes: an updated analysis with a focus on low-and middle-income countries. Int J Hyg Environ Health 222(5):765–777
91. Qiao Y (2018) Preparation, characterization, and evaluation of photocatalytic properties of a novel $NaNbO_3/Bi_2WO6$ heterostructure photocatalyst for water treatment. Université d'Ottawa/University of Ottawa
92. Qin L, He L, Yang W, Lin A (2020) Preparation of a novel iron-based biochar composite for removal of hexavalent chromium in water. Environ Sci Pollut Res 1–13
93. Querido MM, Aguiar L, Neves P, Pereira CC, Teixeira JP (2019) Self-disinfecting surfaces and infection control. Colloids Surf, B 178:8–21
94. Rai PK, Lee J, Brown RJ, Kim K-H (2020) Environmental fate, ecotoxicity biomarkers, and potential health effects of micro-and nano-scale plastic contamination. J Hazardous Mater 123910
95. Rajani A, Sunitha E, Kavitha A, Pankaja D, Anitha S (2017) Water quality monitoring in Fox Sagar Lake, Jeedimetla, Hyderabad, Telangana
96. Ramesh C, Tyagi P, Bhattacharyya B, Husale S, Maurya K, Kumar MS, Kushvaha S (2019) Laser molecular beam epitaxy growth of porous GaN nanocolumn and nanowall network on sapphire (0001) for high responsivity ultraviolet photodetectors. J Alloy Compd 770:572–581
97. Rebosura Jr M, Salehin S, Pikaar I, Kulandaivelu J, Jiang G, Keller J, Yuan Z (2020) Effects of in-sewer dosing of iron-rich drinking water sludge on wastewater collection and treatment systems. Water Res 171: 115396
98. Ricart S, Rico AM (2019) Assessing technical and social driving factors of water reuse in agriculture: a review on risks, regulation and the yuck factor. Agric Water Manag 217:426–439
99. Roselló-Márquez G, Fernández-Domene RM, Sánchez-Tovar R, García-Carrión S, Lucas-Granados B, Garcia-Anton J (2019) Photoelectrocatalyzed degradation of a pesticides mixture solution (chlorfenvinphos and bromacil) by WO3 nanosheets. Sci Total Environ 674:88–95
100. Rossi T, Silva P, De Moura L, Araújo M, Brito J, Freeman H (2017) Waste from eucalyptus wood steaming as a natural dye source for textile fibers. J Cleaner Prod 143:303–310
101. Rozman U, Duh D, Cimerman M, Turk SŠ (2020) Hospital wastewater effluent: hot spot for antibiotic resistant bacteria. J Water Sanitation Hygiene Develop
102. Rtimi S, Dionysiou DD, Pillai SC, Kiwi J (2019) Advances in catalytic/photocatalytic bacterial inactivation by nano Ag and Cu coated surfaces and medical devices. Appl Catal B 240:291–318
103. Sandoval MA, Fuentes R, Thiam A, Salazar R (2020) Arsenic and fluoride removal by electrocoagulation process: a general review. Sci Total Environ 753:142108
104. Sankhla MS, Kumari M, Sharma K, Kushwah RS, Kumar R (2018) Water contamination through pesticide & their toxic effect on human health. Int J Res Appl Sci Eng Technol (IJRASET) 6(1):967–970
105. Schumann M, Brinker A (2020) Understanding and managing suspended solids in intensive salmonid aquaculture: a review. Rev Aquacult
106. Shafiei F, Ashnagar A, Ghavami-Lahiji M, Najafi F, Amin Marashi SM (2018) Evaluation of antibacterial properties of dental adhesives containing metal nanoparticles. J Dental Biomater 5(1):510–519
107. Shao J, Wang B, Bartels CJ, Bronkhorst EM, Jansen JA, Walboomers XF, Yang F (2018) Chitosan-based sleeves loaded with silver and chlorhexidine in a percutaneous rabbit tibia model with a repeated bacterial challenge. Acta Biomater 82:102–110
108. Sigdel B (2017) Water quality measuring station: pH, turbidity and temperature measurement
109. Smith K, Liu S, Liu Y, Guo S (2018) Can China reduce energy for water? A review of energy for urban water supply and wastewater treatment and suggestions for change. Renew Sustain Energy Rev 91:41–58

110. Sohrabi MR, Khavaran A, Shariati S, Shariati S (2017) Removal of Carmoisine edible dye by Fenton and photo Fenton processes using Taguchi orthogonal array design. Arab J Chem 10:S3523–S3531
111. Solis-Casados D, Escobar-Alarcón L, Natividad R, Romero R (2018) Advanced oxidation processes II: removal of pharmaceuticals by photocatalysis. Ecopharmacovigilance. Springer, pp 143–155
112. Soltani T, Lee B-K (2017) Enhanced formation of sulfate radicals by metal-doped $BiFeO_3$ under visible light for improving photo-Fenton catalytic degradation of 2-chlorophenol. Chem Eng J 313:1258–1268
113. Stalter D, Dutt M, Escher B (2013) Headspace-free setup of in vitro bioassays for the evaluation of volatile disinfection by-products. Chem Res Toxicol 26:1605–1614
114. Su M, Zhang T, Su J, Wang Z, Hu Y, Gao Y, Zhang X (2019) Homogeneous ZnO nanowire arrays pn junction for blue light-emitting diode applications. Opt Express 27(16):A1207–A1215
115. Thirunavukkarasu A, Nithya R, Sivashankar R (2020) A review on the role of nanomaterials in the removal of organic pollutants from wastewater. Rev Environ Sci Biotechnology 1–28
116. Tian B, Qiao YY, Tian YY, Xie KC, Liu Q, Zhou HF (2016) FTIR study on structural changes of different–rank coals caused by single/multiple extraction with cyclohexanone and NMP/CS2 mixed solvent. Fuel Process Technol 154:210–218
117. Tope AO, Olufemi AO, Kehinde AB, Omolola FF (2018) Healthcare waste management practices and risk perception of healthcare workers in private healthcare facilities in an urban community in Nigeria
118. Ustaoğlu F, Tepe Y, Taş B (2020) Assessment of stream quality and health risk in a subtropical Turkey river system: a combined approach using statistical analysis and water quality index. Ecol Indicators 113:105815
119. Velo-Gala I, Lopez-Penalver J, Sánchez-Polo M, Rivera-Utrilla J (2017) Role of activated carbon surface chemistry in its photocatalytic activity and the generation of oxidant radicals under UV or solar radiation. Appl Catal B 207:412–423
120. Vodyanitskii YN (2016) Standards for the contents of heavy metals in soils of some states. Ann Agrarian Sci 14(3):257–263
121. Vorontsov AV (2019) Advancing fenton and photo-fenton water treatment through the catalyst design. J Hazard Mater 372:103–112
122. Wahaab RA, Mahmoud M, van Lier JB (2020) Toward achieving sustainable management of municipal wastewater sludge in Egypt: the current status and future prospective. Renew Sustain Energy Rev 127:109880
123. Wang J, Zhang T, Mei Y, Pan B (2018) Treatment of reverse-osmosis concentrate of printing and dyeing wastewater by electro-oxidation process with controlled oxidation-reduction potential (ORP). Chemosphere 201:621–626
124. (2006) Guidelines for the safe use of wastewater, excreta and greywater in agriculture and aquaculture, vol 1–4. World Health Organization, Geneva
125. WHO (2011) Guidelines for drinking-water quality. World Health Organization, Geneva
126. WHO (2016) World health statistics 2016: monitoring health for the SDGs sustainable development goals. World Health Organization
127. WHO (2019a) Causes of child mortality, 2017. Global Health Observatory (GHO) data. World Health Organization, Geneva. 2018 from http://www.who.int/gho/child_health/mortality/causes/en/accessed May 2019
128. WHO (2019b) Malaria: fact sheet. World Health Organization, Geneva from http://www.who.int/news-room/fact-sheets/detail/malaria,accessedMay2019
129. Wurtsbaugh WA, Paerl HW, Dodds WK (2019) Nutrients, eutrophication and harmful algal blooms along the freshwater to marine continuum. Wiley Interdisciplinary Rev Water 6(5):e1373
130. Xie Y, Peng S, Feng Y, Wu D (2020) Enhanced mineralization of oxalate by highly active and stable Ce (III)-Doped g-C_3N_4 catalyzed ozonation. Chemosphere 239:124612

131. Xu M, McKay G (2017) Removal of heavy metals, lead, cadmium, and zinc, using adsorption processes by cost-effective adsorbents. Adsorption processes for water treatment and purification. Springer, pp 109–138
132. Xu Z, Zhao B, Wang Y, Xiao J, Wang X (2020) Composting process and odor emission varied in windrow and trough composting system under different air humidity conditions. Bioresour Technol 297:122482
133. Xue C, Peng Y, Chen A, Peng L, Luo S (2020) Drastically inhibited nZVI-Fenton oxidation of organic pollutants by cysteine: multiple roles in the nZVI/O_2/hv system. J Colloid Interface Sci 582:22–29
134. Yadav V, Karak T, Singh S, Singh AK, Khare P (2019) Benefits of biochar over other organic amendments: responses for plant productivity (*Pelargonium graveolens* L.) and nitrogen and phosphorus losses. Ind Crops Prod 131:96–105
135. Yang J, Bertram J, Schettgen T, Heitland P, Fischer D, Seidu F, Kaifie A (2020) Arsenic burden in e-waste recycling workers–A cross-sectional study at the Agbogbloshie e-waste recycling site, Ghana. Chemosphere 261:127712
136. Yang K, Yu Z, Luo Y, Zhou X, Shang C (2019) Spatial-temporal variation of lake surface water temperature and its driving factors in Yunnan-Guizhou Plateau. Water Resour Res 55(6):4688–4703
137. Yang ZL, Gao BY, Yue QY, Wang Y (2010) Effect of pH on the coagulation performance of Al-based coagulants and residual aluminum speciation during the treatment of humic acid–kaolin synthetic water. J Hazard Mater 178(1–3):596–603
138. Yim C (2019) Organic sulfur-bearing species as subsurface carbon storage vectors. Science
139. Yuan X, Xu Y, Meng H, Han Y, Wu J, Xu J, Zhang X (2018) Fabrication of ternary polyaniline-graphene oxide-TiO_2 hybrid films with enhanced activity for photoelectrocatalytic hydrogen production. Sep Purif Technol 193:358–367
140. Zanin E, Scapinello J, de Oliveira M, Rambo CL, Franscescon F, Freitas L, Dal Magro J (2017) Adsorption of heavy metals from wastewater graphic industry using clinoptilolite zeolite as adsorbent. Process Saf Environ Prot 105:194–200
141. Zhang J, Fu D, Wang S, Hao R, Xie Y (2019) Photocatalytic removal of chromium (VI) and sulfite using transition metal (Cu, Fe, Zn) doped TiO2 driven by visible light: feasibility, mechanism and kinetics. J Ind Eng Chem 80:23–32
142. Zhang L, Lv P, He Y, Li S, Chen K, Yin S (2020) Purification of chlorine-containing wastewater using solvent extraction. J Clean Prod 273:122863
143. Zhang X, Gao B, Creamer AE, Cao C, Li Y (2017) Adsorption of VOCs onto engineered carbon materials: a review. J Hazard Mater 338:102–123
144. Zhao S, Shen Y, Zhou P, Zhong X, Han C, Zhao Q, Wei D (2019) Design of Au@ WO_3 core–shell structured nanospheres for ppb-level NO_2 sensing. Sens Actuat B Chem 282:917–926
145. Zhao S, Wang Z (2017) A loose nano-filtration membrane prepared by coating HPAN UF membrane with modified PEI for dye reuse and desalination. J Membr Sci 524:214–224
146. Zhijiang C, Ping X, Cong Z, Tingting Z, Jie G, Kongyin Z (2018) Preparation and characterization of a bi-layered nano-filtration membrane from a chitosan hydrogel and bacterial cellulose nanofiber for dye removal. Cellulose 25(9):5123–5137
147. Zhou Y, Lu J, Zhou Y, Liu Y (2019) Recent advances for dyes removal using novel adsorbents: a review. Environ Pollut 252:352–365
148. Zou D, Ke X, Qiu M, Chen X, Fan Y (2018) Design and fabrication of whisker hybrid ceramic membranes with narrow pore size distribution and high permeability via co-sintering process. Ceram Int 44(17):21159–21169

Organic–Inorganic Polymer Hybrids for Water and Wastewater Treatment

Md. Ashiqur Rahman, Md. Lawshan Habib, Adib H. Chisty, Abul K. Mallik, M. Nuruzzaman Khan, Papia Haque, and Mohammed Mizanur Rahman

Abstract Day by day, the need of pure drinking water is increasing. Water bodies are being polluted by humans with numerous pollutants like organic, inorganic micro-pollutants. These pollutants alter the chemical compositions, temperature, and whole ecosystem of water bodies. Chemical pollution can directly affect human body from drinking water. To control water pollution, treatment of water is necessary. Most common treatment methods are typically used physical, chemical, and biological water treatment. Organic–inorganic polymer hybrids are a new advancement in water treatment technologies and are very efficient as an adsorbent, flocculants, coagulant, and also useful for the fabrication of water treatment membrane. The use of photocatalysts is a recent advancement in water treatment technologies. This book chapter discusses the synthesis process of organic–inorganic hybrids and recent advancement in water treatment technologies based on organic–inorganic polymer hybrids, polymer composites, polymer nanocomposites, and photocatalyst. The book chapter also discusses the effect and outcomes of organic–inorganic polymer hybrids as an adsorbent, flocculants, coagulant, and photocatalysts and will also discuss the major challenges and future prospects of organic–inorganic polymer hybrids in water treatment.

Md. A. Rahman · A. H. Chisty
Department of Applied Chemistry and Chemical Engineering, National Institute of Textile Engineering and Research (NITER), Nayarhat, Savar, Dhaka 1350, Bangladesh

Md. L. Habib
Department of Applied Chemistry and Chemical Engineering, Faculty of Engineering, Bangabandhu Sheikh Mujibur Rahman Science & Technology University, Gopalganj 8100, Bangladesh

A. K. Mallik (✉) · M. N. Khan · P. Haque · M. M. Rahman
Department of Applied Chemistry and Chemical Engineering, Faculty of Engineering and Technology, University Dhaka, Dhaka 1000, Bangladesh
e-mail: abulkmallik@du.ac.bd

E. Lichtfouse et al. (eds.), *Inorganic-Organic Composites for Water and Wastewater Treatment*, Environmental Footprints and Eco-design of Products and Processes,
https://doi.org/10.1007/978-981-16-5916-4_2

1 Introduction

Water is the most essential ingredients of our life. Around 98% of the total water in the world is seawater, and owing to the high salinity of seawater, it is unavailable for direct use. About 1.6% of the rest is in the polar ice caps and glaciers. The remaining is located in underground, rivers, ponds, or lakes [81].

In 45 developing countries around the world, women and children usually collect freshwater for household purpose. This is time-consuming and affecting the economy adversely. Freshwater sources are nowadays being polluted frequently due to human and industrial activities. With rapid urbanization and industrialization, we need more freshwater, and also freshwater is becoming polluted due to chemical contamination. Industrial, domestic, and agricultural activities are responsible for water pollution in the case of both ground and surface water [75]. More than 80% of the people of the world are in great scarcity of fresh drinking water [88]. Recently, phenols, heavy metals, dyes, pesticides insecticides, and detergents are polluting surface water. These are released from the industries with many other effluents that are hazardous and harmful for our body [75].

In our aquatic system, water-borne pathogens are increasing day by day at an alarming rate. These pathogens enter into the aquatic ecosystem through sewage, drain water, industrial effluent, and medical waste, and so on. Pathogen and toxic chemical-free water is crucial for human health, and contamination of drinking water from natural calamities is a big concern nowadays. In the coming years, supplying toxic chemical and pollutant-free water will be a big challenge. This indicates the requirement for advanced water treatment technologies [47]. Conventional methods used to remove heavy metals, chemicals, inorganic contaminants, dyes, and many other pollutants are adsorption, photodegradation, membrane filtration, chemical oxidation, electrodialysis, reverse osmosis, and so on. These water treatment methods are costly and sometimes produce secondary pollutants. So, it is a crying need to have low cost and highly efficient process and stable approach of water/wastewater treatment. Among many other water treatment processes, photocatalytic degradation is mostly significant owing to its high efficiency, low cost, and easy operating conditions.

Organic–inorganic polymer hybrids are a brilliant edition of photocatalyst to degrade toxic pollutants and dyes obtained from industrial effluent. Organic–inorganic polymer hybrids were recently found to be efficient in different water/wastewater treatments like adsorption, ion exchange, membrane technology, biological contaminants removal [63]. Here, we will explain the preparation process and recent invention of organic–inorganic polymer hybrids in different tertiary water treatment processes and their prospects in water/wastewater treatment.

2 Types of Pollutants in Water Bodies

A well-defined knowledge and awareness about the nature, amount, and sources of pollutants present in water or wastewater are a mandatory to find a best way of water or wastewater treatment. All pollutants are not always harmful to us and the environment. This depends on their quantitative amount that is present in water bodies. Some pollutants are present in the water below their permissible limits and some above their permissible limits. We have to remove or destroy those pollutants that are present in water bodies above their permissible limits. Types of pollutants depend upon their source of release and chemical nature. Pollutants may be classified as organic, inorganic, and biological. Highly toxic and carcinogenic inorganic pollutants are heavy metals and also metals present in water as a salt of nitrates, sulfates, phosphates, fluorides, chlorides, and oxalates. These inorganic compounds are very harmful to our ecosystem and human health. Water bodies are being polluted with insecticides, herbicides, fungicides, hydrocarbons, polymers, phenolic compounds, and toxic aromatics. Pollutants from the textile industry are a big concern for water pollution. Textile dyes, effluent, textile pre-treatment chemicals are nowadays polluting water bodies drastically. Aldehydes, ketones, alcoholic compounds, and medical waste are also responsible for water pollution. Last year, in Russia, microbes from dead bodies found in polar ice caps water indicate a severe catastrophic situation in the near future [15]. Bacteria, fungi, algae, plankton, and viruses are responsible for the biological pollution of water and result in severe water-borne diseases [25].

3 Water Treatment Technologies

Reduction of water source means increase in freshwater, wastewater treatment, and recycling of wastewater for drinking, household, and industrial purposes. Water treatment technologies are divided grossly into three categories. Removal of physical and chemical contaminants is the initial step of preliminary treatment of water/wastewater. After that, biological treatments are regarded as secondary treatment and followed by some advanced treatment process called tertiary treatment. After these treatments, water can be used for drinking, industrial, medicine, etc. The water becomes nearly 99% pollutant-free after the three-stage treatment process and ready for safe use in different specific uses in our daily life. Though, the invention of various technologies for water treatment cost, sustainability, the energy requirement is a big concern. Technological advancement with low cost and energy requirements in material development for the different process is a vital need nowadays. Organic–inorganic polymer hybrids are efficient in advanced water treatment technologies. Different types of organic–inorganic polymer hybrids and their production and application in different water treatment processes will be discussed later in this chapter.

A complete flowchart of these processes to produce good quality pollutant-free safe water is given below

3.1 Primary Water Treatment Technologies

3.1.1 Screening, Filtration, and Centrifugal Separation

The best way of removing waste like cloth, paper, other solids during wastewater treatment is screening. The very first step in water treatment is screening. According to the size of solid waste, different types of screens with required pores are used for solid waste separation. A medium with fine pores is used for the filtration process with a setup of 0.1–0.5 mm pore size. Suspended solids, greases, and oils, bacteria, etc. are removed by filtration. Water obtained after filtration is utilized for further tertiary treatments like adsorption, ion exchange, membrane separation. Non-colloidal solids are usually separated by centrifugal separation. Wastewater rotation is done by rotating devices with different speeds to separate the waste and discharged finally. Suspended solids separation is directly connected to their amount-to-volume ratio. On the other hand, amount of waste separated from wastewater also depends on the speed of the centrifugal machine [32].

3.1.2 Sedimentation and Gravitational Separation

Gravity separation and sedimentation techniques are very useful to separate solid waste, grits, and slits. Water is allowed to settle for a definite time span, and owing to gravitational force, the solid waste will be settled down. The time required for the settling of the waste is dependent to the motion and the density of waste in water. Sometimes, alums are used to supplement the process. Nearly, 60% of solid waste can be separated from wastewater by sedimentation and gravity separation. This is a very usual process in water treatment. Industrial effluents of pulp paper and refinery industries are separated from wastewater by this separation method. Water treated by these processes is applicable to be used ion exchange process and membrane technology [29].

3.1.3 Coagulation

Sedimentation and gravity separation are not useful for the separation of some suspended solids. These non-settable solids are usually present in water in colloidal form. To separate these colloidal suspended solids, chemical treatment is required. Colloidal particle possesses charge on their surface. To stabilize that particle alum, starch or ferric chloride-activated silica and aluminum salts are mostly used as coagulating agents. Due to the high cost, cationic and anionic polymers are not usual in use. pH, temperature, and contact times are very crucial for the coagulation process. Biological wastes can also be separated by using special types of coagulants [46].

3.1.4 Flotation

In traditional water/wastewater treatment project, floatation is an important technique to separate oils, grease, and biological solids, etc. through air or gas flow. Physical adsorption is the chemistry behind this kind of separation process of waste from wastewater. Solids adhere to the surface of gas or air by agglomeration and then skimmed off from the water surface easily by mechanical means. High-pressure air or gas is passed through the wastewater for better separation. The floatation process is an effective process for recycling purpose of wastewater [82].

3.2 Secondary Water Treatment Technologies

To remove soluble and insoluble pollutants, application of microbes is a biological route called secondary water treatment. Microbes are usually bacterial or fungal strains. These microbe-containing water circulation into a reactor with wastewater results in the decomposition product of organic waste. Sometimes, the organic waste results in some other products like alcohol, glucose, and nitrate. On the other hand, these microbes are very useful for the detoxification of toxic wastes in water. The biological treatment of water is divided into two groups. One is an aerobic process, and the other one is an anaerobic process.

3.2.1 Aerobic Process

For the aerobic decomposition of organic waste, availability of dissolved oxygen in wastewater is a must. The decomposition occurs due to aerobic and facultative bacteria. The rate of decomposition is completely dependent on the amount of available oxygen, retention period, temperature, and also on bacteria. To increase bacterial growth followed by biological oxidation rate, some chemicals can be added to enhance the decomposition of organic wastes of wastewater. By applying this technique, solid organic wastes, volatile organics, and nitrate, etc. can be removed effectively. In the aerobic treatment of water, a huge quantity of waste obtained after the treatment is a disadvantage. The further processing of the waste output of the process is costly. But, more than 85% of organic waste can be decomposed through the technique. This is done by a trickling filter or activated sludge process or oxidation pond or lagoons [26]. A general reaction scheme in presence of bacteria is given below

$$\text{Organic pollutants} + O\text{xygen} \rightarrow \text{Decomposition product} \quad (1)$$

3.2.2 Anaerobic Process

Purification occurs in absence of free dissolved oxygen in wastewater, and it is called anaerobic degradation of organic waste of water/wastewater. Anaerobic and facultative bacteria decompose organic pollutants into simpler and small-chain organic molecules containing nitrogen, carbon, and sulfur. Some volatile substances like ammonia, methane, nitrogen, and hydrogen sulfide evolve from wastewater due to decomposition during the process. The method reduces a load of organic waste from wastewater [68].

A general reaction scheme of anaerobic decomposition in presence of bacteria is given below:

$$\text{Organic pollutants} \rightarrow \text{Decomposition product} \tag{2}$$

3.3 *Tertiary Water Treatment Methods*

The ultimate stage of water/wastewater treatment technologies to produce safe water for our daily lives is tertiary water treatment process. The techniques used in these treatment processes are adsorption, ion exchange, oxidation, photodegradation, membrane filtration, reverse osmosis, electrodialysis, distillation, and solvent extraction. In this chapter, we are going to focus on some selected techniques in which organic–inorganic polymer hybrids are used in water/wastewater treatment afterward [32].

3.3.1 Adsorption

Adsorption is a surface phenomena which can be defined as the increase in concentration of any atoms, molecules, or ions on the surface of adsorbent. The rate of adsorption is controlled by temperature, time, and diameter of the adsorbate. Recently, organic–inorganic polymer hybrids have drawn a great attention in water treatment as adsorbents in case of removing heavy metals and organic pollutants like dyes. In industries, pollutants removal from water/wastewater is done by columns or contractors fully loaded with adsorbents. The adsorption process is a batch process and then fixing the condition turns out into the column or contractor's process. Regeneration of columns or contractors is problematic in the adsorption process [30, 31, 72]. The management of adsorption waste is also a matter of concern for scientists.

3.3.2 Distillation

Distillation is a separation technique. In the distillation process, two or multi-component mixture can be separated according to their boiling point. In distillation process, water is purified by boiling it at its boiling point. The water becomes vapor and then cooled as pure water. In this process, the pollutants remain in the distillation tower. The wastewater becomes free from all volatile organics and other impurities. The process can eliminate 99% impurities. According to the water quality and quantity required, various types of boilers with a definite size can be used for distillation. Distilled water is required for laboratory and medical applications. Desalination of seawater is another most important application of distillation [67].

3.3.3 Oxidation

Chemicals like potassium permanganate, potassium dichromate, hydrogen peroxide, chlorine are the most commonly used oxidizing agents in the oxidation process of water/wastewater treatment. Organic chemical waste is oxidized into carbon dioxide, water, and some other small organic compounds like alcohols, aldehydes, ketones, and carboxylic acids. These small organic compounds easily degrade in presence of bacteria. pH, temperature, and catalyst based on organic–inorganic polymer hybrids are very crucial for the oxidation of pollutants. Following these methods, dyes, phenols, ammonia, and many other compounds can be eradicated from water/wastewater [6]. Advanced oxidation process is applicable when single oxidants are not enough for oxidation. In an advanced oxidation process, more than one oxidation applied simultaneously increases the amounts of free radicals in water/wastewater for better oxidation [12, 14]. The most common advanced oxidation process includes oxidation with Fenton's reagents [6], ultraviolet photolysis, O_3, ultraviolet/H_2O_2 [22], etc. Photocatalysis is one of the prominent oxidation processes in organic pollutants degradation. Solar light or UV excites an electron of outer orbital to produce hydroxyl free radicals for oxidation. To degrade organic waste, various metal oxides like TiO_2, ZnO, ZrO_2 with organic polymer composites are very efficient photocatalyst in wastewater treatment [52].

3.3.4 Electrodialysis

Sometimes, water-soluble ions are considered pollutants in water/wastewater treatment. To remove these water-soluble ions, the water is passed through the ion-selective membrane (semi-permeable) under electric current [84]. The membrane is made of ion-selective ion exchange or organic–inorganic polymer hybrid material. The ion exchanger may be cation or anion exchanger, and the exchanger permits cations or anions, respectively. Two electrodes with different voltage are applied to this process in batch or continuous mode. To acquire the desired desalination, the membrane may be applied in series connection. The efficiency of the process is dependent to the electric charge of the pollutants, membranes, water flow rate, fouling and scaling, and several membranes used in series connection. Water source reduction is another application of the technique. Applying electrodialysis technique, more than 90% dissolved solids can be separated from wastewater. Fouling of membrane in reverse osmosis can be solved by increasing the flux and applying of carbon nanotubes in hybrid membranes [30, 31].

3.3.5 Ion Exchange Process

Heavy metal ions in wastewater are a big concern in the effluent treatment plant of any industry. To remove the toxic ions from water, ion exchanger can be used which will exchange the toxic ion with non-toxic metal ion from a material called exchanger. Ion exchangers have naturally obtained resin or synthetic organic–inorganic polymer hybrids with active sites on their surface. The process is reversible and low energy-consuming process. Organic and inorganic pollutants with low concentration can be removed by this process [19, 98].

4 Advancement in Water Treatment Technologies

Widespread studies have been commenced in recent years to find a new way for a feasible and economically profitable way of water and wastewater treatment technologies. Different methods named coagulation, adsorption, photocatalytic degradation, biological treatment process are now being mostly used to separate toxic chemicals from water/wastewater of different sources [3]. Organic–inorganic polymer hybrids open a new era of this tertiary water treatment technologies around the world.

Disinfection of water to remove pathogens obtained from human activities is usually done by chlorination and ozonation. Though these processes are very efficient but advanced organic–inorganic polymer hybrids, semiconductor photovoltaic cells are very sustainable processes with minimum waste in the water treatment industry. These processes are applied to remove persistent pathogens and toxic metallic ions, organic compounds, and microorganisms from wastewater [12].

Adsorption, an effective process for wastewater treatment to remove toxic pollutants, dyes, and many other pollutants. Recently used most widespread adsorbent in water treatment is activated carbon [69]. Recently, biomaterials, known as bio adsorbent, are used to remove pollutants from polluted water [71]. Magnetic adsorbent technology has drawn sufficient attention in water treatment technology. These processes are not widely used in industries. However, a great effort has been given to remove toxic metals, dyes, and other pollutants from wastewater using adsorbents prepared from organic inorganic polymer hybrids.

The ion exchange process is another advanced technique in water treatment, but ion exchangers prepared from organic and inorganic materials lack in mechanical and chemical strength. Organic ion exchangers degrade in presence of radiation due to high sensitivity toward radiation. So, organic–inorganic hybrid polymer-based ion exchangers are, therefore, very potential for water treatment [43, 76]. Photocatalytic degradation pollutants are another advanced method used in water treatment technologies. This is an amazing way for to destroy hazardous pollutants through ozonolysis, UV radiation, and advanced oxidation [91]. This is a concept based on electron–hole pairs of semiconductors representing widely used low cost, easier way to degrade detergents, dye, pesticides, and easily evaporating organic compounds. Photocatalyst made or organic–inorganic hybrids are the new era of water treatment technologies [108].

Membrane technologies are advantageous over many other technologies because of no chemical additives and their non-regenerative properties of spent media. Pressure-driven membrane, gas separation membrane, electrochemical membranes are mostly used nowadays [86].

5 Organic–inorganic Hybrid Polymer

Organic–inorganic polymers hybrids are flexible, lightweight, and versatile materials in material science. Organic–inorganic polymer hybrids are made of polymers with various inorganic materials, and a variety of these synthetic materials are used in different fields like water treatment, medical, and electronics. During the synthesis of the material, the selection of the polymer is mostly important for the stability and mechanical properties. Sometimes, incorporation of nanoparticles of inorganic material with polymer is very crucial for the required characteristics of the hybrid materials. So nowadays, organic–inorganic nanohybrid materials or nanocomposites are widely used in water treatment processes to eradicate the problem arising due to pollutants in water. Now, in the below sections, the methods available to fabricate organic–inorganic polymer hybrids will be discussed [45].

5.1 Fabrication of Organic–inorganic Polymer Hybrids

To fabricate organic–inorganic hybrid material, different established methods are available [55]. Blending, sol–gel synthesis, and emulsion polymerization are mostly used methods for hybrid material synthesis. Metallosupramolecular polymerization, electrochemical synthesis, microwave irradiation are some commonly used techniques for the synthesis of the material.

5.1.1 Blending

The simplest method to synthesize polymer-inorganic hybrid material is solution blending. First of all, a solution is prepared from the polymer and then the inorganic material is incorporated and well-mixed for a homogeneous mixture with mechanical stirring or ultrasound. After that, evaporation of the solvent of polymer solution results in the final products. In the case of inorganic nanoparticles, agglomeration of nanoparticles is a drawback of this low-cost synthetic process. To avoid this problem, the nanoparticles are firstly functionalized with other organic molecules and then introduced in the polymer in liquid phase. In the case of melt blending, the polymer is firstly melted with heat treatment and then the inorganic particles are added to the polymer melt followed by mechanical stirring or ultrasonication. This melt blending is appropriate for nanohybrid synthesis.

Solid-state organic–inorganic hybrid material synthesis is advantageous over solution blending as there is no need for an organic solvent. Thus, the process is not responsible for environmental pollution due to the disposal of the solvent. High loading of inorganic materials in melt or solution of polymer results in high viscosity. To avoid this problem, powder blending is mostly used methods. Polymer blended with silica [61] and TiO_2 [64] is used in water treatment.

5.1.2 Sol–Gel Synthesis

Fabrication of organic–inorganic polymer hybrids through sol–gel technique is an old method, but recently, it is an emerging field for functionalized and structurally advanced hybrid material synthesis. Materials combining inorganic solids with organic polymer at the nanoscale are promising in optical, electrical, catalytic, water treatment, and other application [66]. Figure 1 shows a typical sol–gel method to prepare an organic–inorganic polymer hybrid for heavy metal removal.

In an organic polymer network via sol–gel method, incorporation of silica, alumina, vanadia, titania is most feasible. However, studies showed that silica-based materials are mostly used just because of its highly stable Si–O bond and cheap raw materials. Organic (polymer)–inorganic-based sol–gel hybrids are classified as—

1. Class I
   - Physical interactions: H-bond, ionic bond, etc.
2. Class II
   - Chemical interaction between polymer and inorganic components.

5.1.3 Emulsion Polymerization

Most common heterogeneous polymerization techniques are emulsion, dispersion, micro-, and mini-emulsion [37]. Firstly, to create a nucleation center for polymer

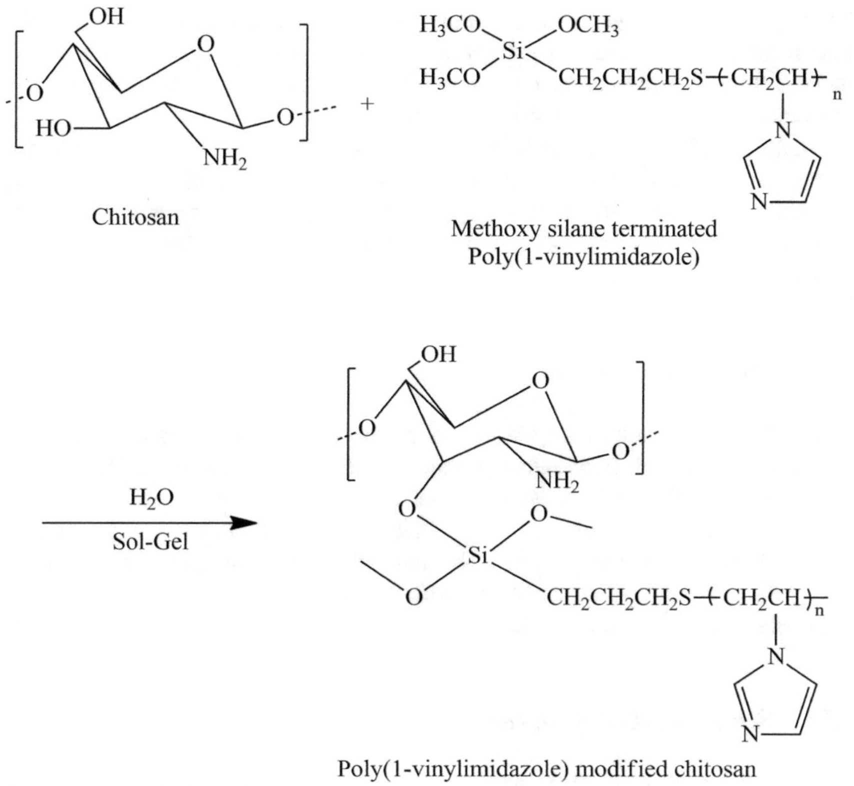

Fig. 1 Preparation of poly(1-vinylimidazole)-modified chitosan composite for heavy metal adsorption from wastewater through sol–gel synthesis. The figure is adapted with permission from the original article of [38] (Copyright 2021 Elsevier)

particle formation, surfactant with hydrophobic monomer is dispersed into a solution. Polymerization radicals will aggregate the micelles, but homogeneous nucleation will result in homogeneous oligomers leading to poor morphological polymers. The mini-emulsion technique is an alternative solution to avoid homogeneous nucleation. In the case of the mini-emulsion technique through anionic, cationic, and polycondensation reactions, polymeric nanoparticles can also be prepared with a free-radical polymer. Applying these methods, polymer inorganic nanocomposites can be prepared with controlled size, shape, and dimensions. Magnetoactive, clay-based, titania-based nanocomposite can be prepared through miniemulsion technique [23, 95].

5.1.4 Metallosupramolecular and Coordination Techniques

In metallosupramolecular approaches, metal incorporates into a synthetic organic polymer backbone. The chemistry behind these approaches is ionic, hydrogen bonding, and metal–ligand coordination complex-based interaction. This selective interaction can be utilized for inorganic–organic polymer hybrid synthesis. Hybrid material from organic polymer and transition-metal element, main group element, and lanthanides are most promising nowadays [96].

Straight-chain polymer-metal ion composite material is a simple method of capturing a metal in between two small polymer molecules with ligating capacity. Recently, branched [103] network structure and cyclic structures [105] are the most popular way of metallosupermolecular synthesis of inorganic–organic polymer hybrid.

5.1.5 Photopolymerization

A suitable photoinitiator firstly results in free radicals due to the application of UV lights on it. Then, the free radical propagates the polymer formation. The polymer formation takes place in between inorganic nanofillers and organic polymer, and the final product is a nanocomposite [99]. Silica, titania, alumina, clays with polymers like acrylates, methacrylates, epoxy, and vinyl polymer are most likely used materials for hybrid material synthesis. Metal oxide nanoparticles are promising inorganic materials for hybrid material synthesis.

5.1.6 Electrochemical Synthesis

Electroactive monomers are utilized for the synthesis of polymer-based nanocomposite through the electrochemical polymerization process. Common monomers are thiophene, aniline, and pyrrole. Direct deposition of polymer-based nanocomposite on electrode surface results in the formation of hybrid material. Electrodeposition of 3-thiophene-acetic acid with magnetite results in a magneto-responsive polymer-based electrode for electrocatalysis [41], and WO_3 nanoporous film with polyaniline electrodeposition is another example of an electrochemical synthesis of metal oxide-polymer-based hybrid material [40].

5.1.7 Surface Grafting

In the case of polymer-based hybrids, the enhancement of specific physical chemical properties is very crucial for the physical or chemical bonding between the organic and inorganic component of the hybrid material. There are three grafting strategies to incorporate inorganic portion with organic polymer, and finally, it will result in the improved interfacial surface. The methods are:

- Grafting to
- Grafting from
- Grafting through [7].

6 Recent Advancement in Water/Wastewater Treatment Technologies

Decontamination becomes a prime requirement when water gets polluted. The most prominent purification technique should be selected to attain the decontamination objectives. An effective decontamination process comprises of five successive steps. During decontamination, polluted water is allowed to undergo preliminary treatment followed by the purification techniques of primary, secondary, and tertiary stages. Generally, both the preliminary and primary stages of water treatment are assembled considering the prevailing conditions of pre-treatment techniques [2, 16].

As a whole, traditional wastewater treatment comprises of an amalgamation of various chemical, biological, and physical methods and procedures to eliminate colloidal and organic impurities, metal impurities, and organic substances, as nutrients, present in several discharges. Various techniques including coagulation and flocculation, adsorption on activated carbon, precipitation, filtration, and biodegradation as traditional processes evaporation, oxidation, solvent extraction, membrane separation, ion exchange, incineration, etc. as established recovery methods and adsorption onto non-conventional solids, biosorption, nanofiltration, advanced oxidation, etc. as emerging removal actions have recently been found to be applied efficiently as decontamination process [17, 70].

6.1 Organic–Inorganic Polymer Hybrids as Coagulant/Flocculants

Coagulation can be defined as the process to overcome the obstruction by repulsion between the elements along with the modification of ionic character. In case of separating a mixture of solid and liquid, both the flocculation and coagulation processes are considered significant in the treatment of wastewater and dewatering of sludge in various industries [50, 109]. Coagulation of impurities, both colloidal and dissolved, in effluent can be clarified following the concept known as Derjaguin–Landau–Verwey–Overbeek theory (DLVO theory). Incorporation of inorganic coagulants, for instance, substances based on iron and aluminum frequently show a key role in counterbalancing the charges present on the surface of the particles in suspended or colloidal arrangements that facilitate the combination and settling of particles owing to the force of gravity resulting in electrical double layer compression [57]. Alternatively, flocculation can be defined as the addition of chemical reagents that can form floc, following the coagulation process, to combine solids of both the suspended and colloidal nature. The process acts a chief character in removing impurities present in the aqueous environments through proper linking of the collected flocs resulting in large clumps in the company of polymeric materials [78].

Numerous materials, for example, flocculants of organic nature, coagulants of inorganic nature, and fused materials of hybrid character have been found to be modified in current centuries for both the flocculation and coagulation methods [60]. Though a varieties of materials has successfully been developed in eliminating pollutants from wastewater, an improvement in their performances is still in need. The successive increase of demands for both the resourceful and operational constituents in the management of wastewater has introduced the growth of fused materials having hybrid nature in case of flocculation and coagulation. Therefore, fused substances of hybrid nature, thus, have developed promisingly as innovative materials having remarkable prospective in the treatment of wastewater owing to their functional superiority compared to conventional coagulants together with its poor cost than the flocculants of organic nature [94]. Fused materials, utilized during the treatment of wastewater, are the substances usually attained through the incorporation of potential constituents into the matrix to improve combining power. It is consistent with the growth of functionalities or constituents into the parent species having its ability to strengthen the aggregating power [85]. Owing to the superior estrogenic character of the constituents in a single substance, fused substance usually gives enhanced performance compared to distinct components [101]. Following the same, hybrid materials with combined useful constituents, can be a suitable substitute for the treatment of contaminated water compared to individual coagulant/flocculant.

The use of mineral salts usually provides a drawback during coagulation, whereas the coagulation capacity is lower than that of IPCs (refers to pre hydrolyzed coagulants) and organic flocculants when they are allowed to be applied separately. In order to deal with this restriction, polymeric flocculants of organic origin, i.e., polydimethyldiallylammonium chloride and polyacrylamide are frequently utilized to fabricate the hybrids comprising of both the organic and inorganic materials. Both the polyacrylamide and polydimethyldiallylammonium chloride represent general properties of higher molecular weight and higher water solubility. Inorganic coagulants and IPCs such as aluminum hydroxide [80], calcium chloride [49], ferric chloride [49], magnesium chloride, and magnesium hydroxide [48], polyferric chloride, polyaluminum chloride have recently been found to combine with polyacrylamide resulting in fused materials comprising of both inorganic and organic constituents. Through the successful incorporation of polyacrylamide, the linking process can also be modified to increase the combining capability consecutively [85]. On the other hand, polydimethyldiallylammonium chloride has been reported to not only efficient in eliminating the impurities present in the contaminated water but also capable of minimizing the formulation of chloroform in water which is in need of further treatment. Thus, polydimethyldiallylammonium chloride has been found to be nominated to combine with numerous substances of inorganic nature, i.e., $Al_2(SO_4)_3$ [106], $FeCl_3$ [90], $FeSO_4$ [53], PFS-$FeSO_4$ [53], PAC [54] while improving the coagulation/flocculation efficacy and reducing the dosage of inorganic substance as well [24].

6.2 Organic–Inorganic Polymer Hybrids as Adsorbent

A range of hazardous chemical species from both the organic and inorganic origin is frequently allowed to discharge into the atmosphere as commercial wastes resulting in severe atmospheric pollution. Heavy metals (for instance, Pb, Hg, Co, Cu, Ni, and Pd) are considered as the regular components of the earth surface and exist in the atmosphere as a consequence of enduring besides destruction of parental stuns [73]. Besides natural sources, heavy metals get involved in ecosystems through wastewaters by man-made sources including the fabrication of alloys, mining, metal extraction, tanning, plating, and fertilizer-containing metal ions [9]. Such hazardous ions of metallic origin have been found to deteriorate various water resources and drinking water even at a lower concentration. Therefore, it is essential to eliminate these contaminant from commercial discharges for their successive non-toxic release.

Numerous methods, for example, precipitation which involves deposition of solid contaminants [21], dissolution utilizing appropriate solvents [51] electrochemical and chemical methods [20], ultrafiltration and reverse osmosis, ion exchange technique, flotation and coagulation, etc. have been found to be well-known for the elimination of carcinogenic metallic ions present in commercial effluents and polluted water [28, 56]. Nonetheless, maximum of these techniques are objectionable due to the proper dumping of sludge, lower efficacy, excessive cost, and inappropriateness to a range of contaminants as well [5].

Adsorption is an eminent separation technique and acknowledged as one of the effective and profitable approaches in purifying polluted water. Furthermore, most of the purifier, i.e., adsorbents can easily be recovered following an appropriate releasing technique [65]. At present, clays, zeolites, activated carbon, and residues of agriculture are utilized for the elimination of hazardous metallic ions [87, 104]. However, the foremost drawbacks of such adsorbents are enlisted as their relatively weak bonding with metallic ions, lower absorption capability, complications regarding the separation, and regeneration from water.

To overcome these limits, organic–inorganic hybrid polymers are used most prominently for the elimination of hazardous species in recent times [92, 93]. Among the chemical species, the practical distinction of organic entities is allowed to combine with the benefits of the heat stable and vigorous inorganic substrates which in turn results in durable binding attractions to the specific ions of metallic character.

In various studies, researchers have manufactured ionic polymers that are imprinted (IIPs) for the successful adsorption of metal ions [66, 97]. IIPs were found to be manufactured by means of successful modification of a supporting exterior (for instance, silica gel) utilizing a chelating group of organic origin and consequently printed with metallic ions used as a model or subsequent co-condensation of monomers comprising of functionalic entities (when metallic ions are present) following the co-condensation via sol–gel methods and lastly discharging ions of metallic character from the fabricated compounds. For instance, thiocyanate- and derivatized monomers were reported to produce Cd^{2+}-printed long-chained polymers [73]. Such prepared compounds showed promising adsorption capacity together

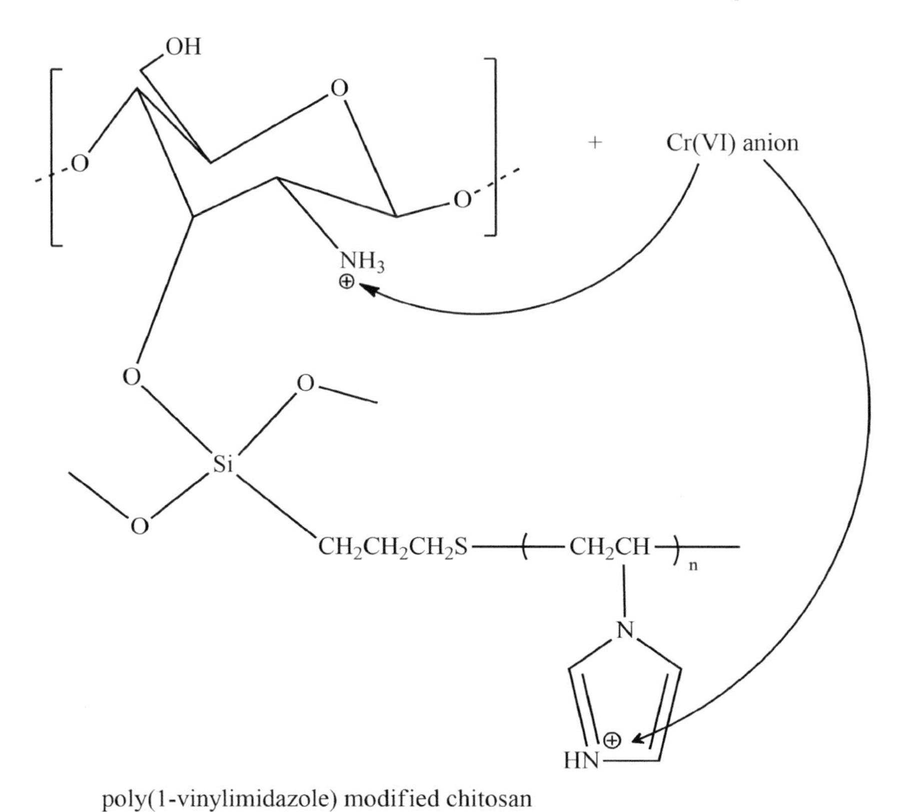

Fig. 2 Mechanism of chromium (IV) adsorption with poly(1-vinylimidazole) modified chitosan. The figure is adapted with permission from the original article of [38] (Copyright 2021 Elsevier)

with higher selectivity toward templating ions. Chitosan and poly(1-vinylimidazole) modified with trimethoxysilyl group is a remarkable advancement in chromium adsorption from wastewater. The mechanism of chromium adsorption by this composite is shown in the following Fig. 2.

Chitosan-functionalized bentonite, perlite, clinoptilolite, alumina, montmorillonite, and calcium alginate were also assessed to be used for the adsorption of carcinogenic metals from water [73] (Table 1).

6.3 *Organic–Inorganic Polymer Hybrids as Photocatalyst*

Among the several advanced procedures, the photocatalytic process has drawn significant consideration as an efficient procedure for the damage of organic impurities in water. Such a procedure is an effective alternative for the elimination of hazardous and carcinogenic contaminants in wastewater. The system utilizes light

Table 1 Organic–inorganic polymer hybrids as adsorbent for the removal of dyes and heavy metal

Serial number	Adsorbent	Adsorbate	Adsorption capacity (mg/g)	Reference
01	Natural clay and aminopropyl-triethoxisilane	Acid red 1	364.1	[83]
02	Clay and aminopropyl-triethoxisilane	Acid green 25	397	[83]
03	Di-o-benzo-p-xylyl-28-crown-8-ether anchored mesoporous silica	Cesium	97.63	[4]
04	Poly(methacrylate)/silica	Methylene blue	56.625 to 91.324	[39]
05	Sodium alginate—SiO_2	Methylene blue	148.23	[36]
06	Carboxymethyl cellulose—TiO_2	Methylene blue	48.73	[107]
07	Poly(acrylic acid)$MnFe_2O_4$	Methylene blue	53.50	[89]
08	Poly-(3-glycidyloxypropyltrimethox-ysilane)/titanium isopropoxide	Pb^{2+}	199.0	[8]
09	Poly-(3-glycidyloxypropyltrimethox-ysilane)/titanium isopropoxide	Cu^{2+}	42.79	[8]
10	Sepiolite–cellulose	Malachite green	62.34	[42]

having wavelength energy (bandgap energy) which in turn stimulates the semiconductors producing a pair of electron–hole. Hence, the photocatalytic method is very modest, cheap and can be applied for a varieties of impurities (including volatile organic compounds, dyes, detergents, and pesticides) [11, 33].

Owing to the adaptability with long-chained polymers (lyophobic or lyophilic nature), TiO_2/cellulose nanocomposites are marked as fore-runners in numerous uses that include both the self-cleaning materials and air purification filters [13]. The utilization of cellulosic sheath as an auxiliary photocatalyst nanoparticle is capable of overcoming the difficulties in gathering and removing photocatalyst suspension that are present in water once the photocatalytic treatment is done successfully. Hence, it lessens the adulteration owing to the presence of photocatalyst in water that is undergoing treatment. Nano-sized cellulosic fiber (designated as CNF) can also be utilized, as a potential matrix, in case of the suspension comprising of photocatalytic particles owing to the required optical and tensile characters [77].

During the photodegradation process, ophthalmic characters of the cellulosic film are significantly imperious for the maximization of titanium dioxide, exposed by a source that emits light, inside the matrix of films. Such transference character of the cellulosic film results in the improvement of electron spreading and transmission to the exterior of titanium dioxide consecutively. Therefore, it will upsurge the catalytic performance of titanium dioxide while considering the impurities. An earlier assessment explored the reduction in ophthalmic transference which can further be attributed to the proper distribution of nano-sized filler in the matrix of regenerated cellulose (RC) as a result of prominent interfacial bonding between titanium dioxide and regenerated cellulose [58]. Moreover, Morawski et al. have magnificently manufactured nano-sized composites comprising of cellulose and titanium dioxide with greater absorption of UV via light through the amalgamation of both hexyl-dimethylammonium acetate and dimethyl sulfoxide as a solvent system [59].

6.4 Organic–Inorganic Polymer Hybrids in Membrane Technologies

Over the past recent years, membrane technology has become a dignified separation technology as it works without the addition of chemicals, with relatively low energy use and easy and well-arranged process conductions. Polymer hybrids have added a new dimension to membrane technology.

6.4.1 Classification of Organic–inorganic Nanocomposite Membranes

According to the structure of organic–inorganic nanocomposite membranes, they can be divided into two types 10:

(a) Type (I) in which van der Waals force or hydrogen bonds are involved in between polymer and inorganic phases. In this type of system, no covalent or ionic bonds are present, instead various weak interactions like hydrogen bonding, van der Waals contacts, π-π interactions, or electrostatic forces are present.
(b) Type (II) in which strong chemical bonds, i.e., covalent bonds or Lewis acid–base bonds are involved in polymer and inorganic phases 74.

6.4.2 Processing of Organic–Inorganic Nanocomposite Membranes

Hybrid membranes can be prepared by the following three ways:

i. The sol–gel process.
ii. The phase inversion method or the in situ blending method and
iii. In situ or interfacial polymerization.

(i) Sol–gel Process: To achieve highly homogeneous and controlled morphology of the prepared polymer–inorganic nanocomposite membranes, this method is applied. An enhancement in porosity and thermal stability was observed in polysulfone (PS) with titanium dioxide (TiO_2) composite membrane that was prepared by this method 102. Another inorganic–organic polymer hybrid PEO–$[Si(OCH_3)_3]_2$ was also synthesized by coupling N-[3-(trimethoxysilyl)propyl] ethylenediamine (A-1120) to end-capped PEO-400 and then a number of positively charged membranes were also prepared based on this hybrid 18. Joly et al. prepared a polyimide film by the sol–gel technique, and the effect of silica particles was studied on the gas transport properties 44.

(ii) Phase Inversion Method: Phase inversion method is one of the most common methods that is used in the preparation of polymeric membranes. A composite ultrafiltration membrane was prepared by Yan et al. using poly(vinylidene fluoride) PVDF and alumina (Al_2O_3) materials by the phase inversion method, and they studied the characteristics like the membrane hydrophilicity, porosity, protein retention, and surface morphologies 100. Other studies have been also conducted such as incorporating nano-TiO_2 into regenerated cellulose (RC) [107] polysulfone (PSF)/silica (SiO_2) 1, polyethersulfone (PES) and modified montmorillonite (OMMT) 27, etc.

(iii) In Situ/Interfacial Polymerization: In situ or interfacial polymerization involves the mixing of inorganic nanoparticles with organic monomers followed by the polymerization of the monomers. As interfacial polymerization method is closely related to the production of reverse osmosis membranes, researchers are being attracted toward this technique. In these researches, polyamide is most common 79.

6.5 *Organic–Inorganic Polymer Hybrids in Disinfection of Biological Contaminants*

Organic–inorganic hybrids are mainly the combination of organic materials such as polymeric and carbonaceous materials with inorganic nanomaterials. Generally, Ag, ZnO, TiO_2, and iron oxide nanoparticles have been used as antimicrobial agents for the removal of pathogens from water. Nanotechnology has a lot of potential for water disinfection, and recent reviews and books suggest its importance in water treatment [53]. Considering various issues such as cost, environmental toxicity, and human exposure related to the inorganic materials, organic–inorganic material hybrids such as polymer–Ag, organic–silica, and organic–TiO_2 have been developed as effective materials for water disinfection, particularly for developing nations.

6.5.1 Polymer–Ag Composites

Silver (Ag^+ or nanoparticle) as an antimicrobial agent has a long history of being widely used. Ag^+ has a strong interaction with thiols on enzymes and bacterial cell membranes, causing denaturation and cellular damage, respectively, ultimately leading to cell death. Though it is effective against a broad range of microorganisms, Ag has no significant side effects on the human body when present at sufficiently low levels. For the controlled release of silver that can offer long-lasting antimicrobial efficacy, silver has been impregnated into various materials, including polymers, carbonaceous, and mesoporous materials.

The incorporation of silver into polymers is a promising strategy for the development of materials for water disinfection. The antibacterial efficiency of cylindrical polypropylene water filters coated with a 35.0 nm layer of nano-silver particles (nAg) has been evaluated 34. In another report, nAg-impregnated functionalized carbon nanotubes (multiwall carbon nanotubes [MWNTs]) polymerized with β cyclodextrin (CD) using hexamethylene diisocyanate as the linker has been reported (Uppu et al. 2014). The polymeric nanocomposites (Ag-MWNT-CD) were found to reduce bacterial cell counts in water spiked with E. coli to 94% within 30 min and as low as 0 CFU/mL (colony-forming units per milliliter) within 90 min 35.

Another strategic design of polymer–Ag nanocomposites for the development of environmentally safe water disinfection processes is using the core–shell Ag nanoparticles or nanocomposites with both antibacterial and magnetic properties.

6.5.2 Organic–Silica Composites

A new sand filtration water disinfection technology has been developed by Hayden and coworkers that relies on the antimicrobial properties of hydrophobic polycations (N-hexylated PEI) covalently attached to the sand's surface 62. The applicability of this filter in water disinfection was evaluated both with water spiked with E. coli and with effluents from a wastewater treatment plant.

6.5.3 Organic–Inorganic TiO_2 Composites

Photodegradation of bacteria with organic–TiO_2 composites has the potential for use in water disinfection if the visible light photocatalytic antimicrobial activity can be enhanced.

7 Future Aspects and Major Challenges

Organic–inorganic polymer hybrids show a great prospect and advancement in water/wastewater treatment due to their unparalleled and revolutionizing properties. Here, in this chapter, we have discussed the synthesis of organic–inorganic polymer hybrids and their application in water/wastewater treatment. Organic–inorganic polymer hybrids can act as an efficient photocatalyst, membrane, coagulant or flocculants, adsorbent, and it can also remove organic micropollutants from wastewater. Shortly, advancement of organic–inorganic polymer hybrids synthesis can bring a revolutionary change in water treatment. Ion exchange resin or hybrid polymeric material, the membrane is also promising in many other applications except water treatment. Recently, advancement in membrane technology of organic–inorganic polymer hybrids is very promising for water purification. But, due to the complex synthesis process and cost, sometimes, organic–inorganic polymer hybrids are not feasible for use in water/wastewater treatment. Therefore, improvising the synthesis with different inorganic nanoparticles with various properties like the easy generation of free radical, antimicrobial activity, and high surface to volume ration can be a tremendous improvement in water/wastewater treatment.

8 Conclusion

In summary, organic–inorganic polymer hybrids have drawn the attention of scientists and researchers, and numerous research articles have been published in water/wastewater treatment. High efficiency of organic–inorganic polymer hybrids is evidently proofed in tertiary water treatment. In the case of adsorption, membrane technology, coagulation, ion exchange, biological waste separation organic–inorganic polymer hybrid leads to a safe and creative effective water/wastewater treatment process. Though to obtain a much better efficiency modification, improvisation is necessary during the synthesis and application of inorganic–organic polymer hybrids.

References

1. Ahn J, Chung W.-J, Pinnau I, Guiver MD (2008) Polysulfone/silica nanoparticle mixed-matrix membranes for gas separation. J Membr Sci 314(1–2):123–133
2. Anjaneyulu Y, Chary NS, Raj DSS (2005) Decolourization of industrial effluents–available methods and emerging technologies–a review. Rev Environ Sci Bio/Technol 4(4):245–273
3. Association AWW (1990) Water quality and treatment: a handbook of community water supplies. McGraw-Hill Companies
4. Awual MR, Miyazaki Y, Taguchi T, Shiwaku H, Yaita T (2016) Encapsulation of cesium from contaminated water with highly selective facial organic–inorganic mesoporous hybrid adsorbent. Chem Eng J 291:128–137
5. Barakat MA (2011) New trends in removing heavy metals from industrial wastewater. Arab J Chem 4(4):361–377
6. Bautista P, Mohedano AF, Casas JA, Zazo JA, Rodriguez JJ (2008) An overview of the application of Fenton oxidation to industrial wastewaters treatment. J Chem Technol Biotechnol: Int Res Process, Environ Clean Technol 83(10):1323–1338
7. Beija M, Marty J-D, Destarac M (2011) RAFT/MADIX polymers for the preparation of polymer/inorganic nanohybrids. Prog Polym Sci 36(7):845–886
8. Bozbas SK, Ay U, Kayan A (2013) Novel inorganic–organic hybrid polymers to remove heavy metals from aqueous solution. Desalin Water Treatment 51(37–39):7208–7215
9. Bradl H (2005) Heavy metals in the environment: origin, interaction and remediation. Elsevier
10. Cong H, Radosz M, Towler BF, Shen Y (2007) Polymer–inorganic nanocomposite membranes for gas separation. Sep Purif Technol 55(3):281–291
11. Chen D, Wang K, Xiang D, Zong R, Yao W, Zhu Y (2014) Significantly enhancement of photocatalytic performances via core–shell structure of ZnO@ mpg-C3N4. Appl Catal B 147:554–561
12. Chong MN, Jin B, Chow CW, Saint C (2010) Recent developments in photocatalytic water treatment technology: a review. Water Res 44(10):2997–3027

13. Colmenares JC, Varma RS, Lisowski P (2016) Sustainable hybrid photocatalysts: titania immobilized on carbon materials derived from renewable and biodegradable resources. Green Chem 18(21):5736–5750
14. Comninellis C, Kapalka A, Malato S, Parsons SA, Poulios I, Mantzavinos D (2008) Advanced oxidation processes for water treatment: Advances and trends for R&D. J Chem Technol Biotechnol: Int Res Process, Environ Clean Technol 83(6):769–776
15. Craig RK (2020) Warming oceans, coastal diseases, and climate change public health adaptation. Sea Grant L & Pol'y J 10:3
16. Crini G, Badot P-M (2011) Sorption processes and pollution: Conventional and non-conventional sorbents for pollutant removal from wastemasters. Presses Univ, Franche-Comté
17. Crini G, Lichtfouse E (2019) Advantages and disadvantages of techniques used for wastewater treatment. Environ Chem Lett 17(1):145–155
18. Cuiming W, Tongwen X, & Weihua Y (2003) Fundamental studies of a new hybrid (inorganic–organic) positively charged membrane: Membrane preparation and characterizations. J Membr Sci 216(1–2):269–278
19. Dabrowski A, Hubicki Z, Podkościelny P, Robens E (2004) Selective removal of the heavy metal ions from waters and industrial wastewaters by ion-exchange method. Chemosphere 56(2):91–106
20. Emamjomeh MM, Sivakumar M (2009) Review of pollutants removed by electrocoagulation and electrocoagulation/flotation processes. J Environ Manage 90(5):1663–1679
21. Esalah JO, Weber ME, Vera JH (2000) Removal of lead, cadmium and zinc from aqueous solutions by precipitation with sodium Di-(n-octyl) phosphinate. Can J Chem Eng 78(5):948–954
22. Esplugas S, Bila DM, Krause LGT, Dezotti M (2007) Ozonation and advanced oxidation technologies to remove endocrine disrupting chemicals (EDCs) and pharmaceuticals and personal care products (PPCPs) in water effluents. J Hazard Mater 149(3):631–642
23. Faucheu J, Gauthier C, Chazeau L, Cavaille J-Y, Mellon V, Lami EB (2010) Miniemulsion polymerization for synthesis of structured clay/polymer nanocomposites: short review and recent advances. Polymer 51(1):6–17
24. Gao B-Y, Wang Y, Yue Q-Y, Wei J-C, Li Q (2007) Color removal from simulated dye water and actual textile wastewater using a composite coagulant prepared by ployferric chloride and polydimethyldiallylammonium chloride. Sep Purif Technol 54(2):157–163
25. Gaston V (1979) International regulatory aspects for chemicals, vol I. CRC Press Inc., New York
26. Ge H, Batstone DJ, Keller J (2013) Operating aerobic wastewater treatment at very short sludge ages enables treatment and energy recovery through anaerobic sludge digestion. Water Res 47(17):6546–6557
27. Ghaemi N, Madaeni SS, Alizadeh A, Rajabi H, Daraei P (2011) Preparation, characterization and performance of polyethersulfone/organically modified montmorillonite nanocomposite membranes in removal of pesticides. J Membr Sci 382(1–2):135–147
28. Ghurye G, Clifford D, Tripp A (2004) Iron coagulation and direct microfiltration to remove arsenic from groundwater. J Am Water Works Assoc 96(4):143–152
29. Grimes BA (2012) Population balance model for batch gravity separation of crude oil and water emulsions. Part I: model formulation. J Dispersion Sci Techn 33(4):578–590. https://doi.org/10.1080/01932691.2011.574946
30. Gupta VK, Agarwal S, Saleh TA (2011) Chromium removal by combining the magnetic properties of iron oxide with adsorption properties of carbon nanotubes. Water Res 45(6):2207–2212. https://doi.org/10.1016/j.watres.2011.01.012
31. Gupta VK, Agarwal S, Saleh TA (2011) Synthesis and characterization of alumina-coated carbon nanotubes and their application for lead removal. J Hazard Mater 185(1):17–23
32. Gupta VK, Ali I, Saleh TA, Nayak A, Agarwal S (2012) Chemical treatment technologies for waste-water recycling—an overview. Rsc Adv 2(16):6380–6388
33. Hashimoto K, Irie H, Fujishima A (2005) TiO2 photocatalysis: a historical overview and future prospects. Jpn J Appl Phys 44(12R):8269

34. Heidarpour F, Ghani WWAK, Fakhru'l-Razi A, Sobri S, Heydarpour V, Zargar M, Mozafari MR (2011) Complete removal of pathogenic bacteria from drinking water using nano silver-coated cylindrical polypropylene filters. Clean Technol Environ Policy 13(3):499–507
35. Hiro ME, Pierpont YN, Ko F, Wright TE, Robson MC, Payne WG (2012) Comparative evaluation of silver-containing antimicrobial dressings on in vitro and in vivo processes of wound healing. Eplasty, 12
36. Hosseinzadeh H, Abdi K (2017) Efficient removal of methylene blue using a hybrid organic–inorganic hydrogel nanocomposite adsorbent based on sodium alginate–silicone dioxide. J Inorg Organomet Polym Mater 27(6):1595–1612
37. Hu J, Chen M, Wu L (2011) Organic-inorganic nanocomposites synthesized via miniemulsion polymerization. Polymer Chem 2(4):760–772
38. Islam MN, Khan MN, Mallik AK, Rahman MM (2019) Preparation of bio-inspired trimethoxysilyl group terminated poly(1-vinylimidazole)-modified-chitosan composite for adsorption of chromium (VI) ions. J Hazard Mater 379:120792
39. Jamwal HS, Kumari S, Chauhan GS, Reddy NS, Ahn J-H (2017) Silica-polymer hybrid materials as methylene blue adsorbents. J Environ Chem Eng 5(1):103–113
40. Janaky C, de Tacconi NR, Chanmanee W, Rajeshwar K (2012) Electrodeposited polyaniline in a nanoporous WO3 matrix: an organic/inorganic hybrid exhibiting both p-and n-type photoelectrochemical activity. J Phys Chem C 116(6):4234–4242
41. Janáky C, Visy C, Berkesi O, Tombácz E (2009) Conducting polymer-based electrode with magnetic behavior: electrochemical synthesis of poly (3-thiophene-acetic-acid)/magnetite nanocomposite thin layers. J Phys Chem C 113(4):1352–1358
42. Jiang X, Wang S, Ge L, Lin F, Lu Q, Wang T, Huang B, Lu B (2017) Development of organic–inorganic hybrid beads from sepiolite and cellulose for effective adsorption of malachite green. RSC Adv 7(62):38965–38972
43. Jindal R, Sharma R, Maiti M, Kaur H (2016) In air synthesis of psyllium based organo-inorganic hybrid ion exchanger, its characterization and studies. Int J Sci Eng Manage 1:22–29
44. Joly C, Goizet S, Schrotter JC, Sanchez J, Escoubes M (1997) Sol-gel polyimide-silica composite membrane: Gas transport properties. J Membr Sci 130(1–2):63–74
45. Krasia-Christoforou T (2015) Organic–inorganic polymer hybrids: synthetic strategies and applications. In: Hybrid and hierarchical composite materials, pp 11–63. Springer
46. Kumar V, Othman N, Asharuddin S (2017) Applications of natural coagulants to treat wastewater—a review. MATEC Web of Conf 103:06016
47. Kumar Reddy DH, Lee SM (2012) Water pollution and treatment technologies. J Environ Anal Toxicol, 2:e103
48. Lee KE, Khan I, Morad N, Teng TT, Poh BT (2011) Thermal behavior and morphological properties of novel magnesium salt–polyacrylamide composite polymers. Polym Compos 32(10):1515–1522
49. Lee KE, Teng TT, Morad N, Poh BT, Hong YF (2010) Flocculation of kaolin in water using novel calcium chloride-polyacrylamide (CaCl2-PAM) hybrid polymer. Sep Purif Technol 75(3):346–351
50. Lee KE, Teng TT, Morad N, Poh BT, Mahalingam M (2011) Flocculation activity of novel ferric chloride–polyacrylamide (FeCl3-PAM) hybrid polymer. Desalination 266(1–3):108–113
51. Lertlapwasin R, Bhawawet N, Imyim A, Fuangswasdi S (2010) Ionic liquid extraction of heavy metal ions by 2-aminothiophenol in 1-butyl-3-methylimidazolium hexafluorophosphate and their association constants. Sep Purif Technol 72(1):70–76
52. Li XZ, Liu H, Cheng LF, Tong HJ (2003) Photocatalytic oxidation using a new catalyst TiO2 microsphere for water and wastewater treatment. Environ Sci Technol 37(17):3989–3994
53. Li R, Zheng Y, Peng Z, Peng C (2008) Simulated dyeing wastewater treated by DMDAAC and its composite flocculant. J Central South Univ Sci Technol 39(4):658–664
54. Lu L, Gao BY, Xu CH, Yue QY, Cao BC, Xu SP, Li WW (2007) Municipal wastewater treatment using a composite flocculant made of polyaluminum chloride and polydimethyldiallyammonium chloride. Huan Jing Ke Xue = Huanjing Kexue 28(9):2035–2040

55. Luan J, Wang S, Hu Z, Zhang L (2012) Synthesis techniques, properties and applications of polymer nanocomposites. Curr Org Synth 9(1):114–136
56. Mahmoud A, Hoadley AF (2012) An evaluation of a hybrid ion exchange electrodialysis process in the recovery of heavy metals from simulated dilute industrial wastewater. Water Res 46(10):3364–3376
57. Mensah, J. (2005) Structure formation in dispersed systems. Surf Sci Ser 126:135–216
58. Mohamed MA, Salleh WNW, Jaafar J, Ismail AF, Abd Mutalib M, Sani NAA, Asri S, Ong CS (2016) Physicochemical characteristic of regenerated cellulose/N-doped TiO_2 nanocomposite membrane fabricated from recycled newspaper with photocatalytic activity under UV and visible light irradiation. Chem Eng J 284:202–215
59. Morawski AW, Kusiak-Nejman E, Przepiórski J, Kordala R, Pernak J (2013) Cellulose-TiO_2 nanocomposite with enhanced UV–Vis light absorption. Cellulose 20(3):1293–1300
60. Moussas PA, Zouboulis AI (2009) A new inorganic–organic composite coagulant, consisting of polyferric sulphate (PFS) and polyacrylamide (PAA). Water Res 43(14):3511–3524
61. Olmos D, Rodríguez-Gutiérrez E, González-Benito J (2012) Polymer structure and morphology of low density polyethylene filled with silica nanoparticles. Polym Compos 33(11):2009–2021
62. Onnis-Hayden A, Hsu BB, Klibanov AM, Gu AZ (2011) An antimicrobial polycationic sand filter for water disinfection. Water Sci Technol 63(9):1997–2003
63. Opoku F, Kiarii EM, Govender PP, Mamo MA (2017) Metal oxide polymer nanocomposites in water treatments. In: Descriptive inorganic chemistry researches of metal compounds. IntechOpen
64. Ouyang G, Wang K, Chen XY (2012) TiO_2 nanoparticles modified polydimethylsiloxane with fast response time and increased dielectric constant. J Micromech Microeng 22(7), 074002
65. Pan B, Pan B, Zhang W, Lv L, Zhang Q, Zheng S (2009) Development of polymeric and polymer-based hybrid adsorbents for pollutants removal from waters. Chem Eng J 151(1–3):19–29
66. Pandey S, Mishra SB (2011) Sol–gel derived organic–inorganic hybrid materials: synthesis, characterizations and applications. J Sol-Gel Sci Technol 59(1):73–94
67. Pangarkar BL, Sane MG, Parjane SB, Guddad M (2014) Status of membrane distillation for water and wastewater treatment—A review. Desalin Water Treatment 52(28–30):5199–5218
68. Rajagopal R, Saady NMC, Torrijos M, Thanikal JV, Hung Y-T (2013) Sustainable agro-food industrial wastewater treatment using high rate anaerobic process. Water 5(1):292–311
69. Rao MM, Reddy DK, Venkateswarlu P, Seshaiah K (2009) Removal of mercury from aqueous solutions using activated carbon prepared from agricultural by-product/waste. J Environ Manage 90(1):634–643
70. Rathoure AK, Dhatwalia VK (2016) Toxicity and waste management using bioremediation. Eng Sci Ref
71. Reddy DHK, Ramana DKV, Seshaiah K, Reddy AVR (2011) Biosorption of Ni (II) from aqueous phase by Moringa oleifera bark, a low cost biosorbent. Desalination 268(1–3):150–157
72. Saleh TA, Gupta VK (2012) Column with CNT/magnesium oxide composite for lead (II) removal from water. Environ Sci Pollut Res 19(4):1224–1228
73. Samiey B, Cheng C-H, Wu J (2014) Organic-inorganic hybrid polymers as adsorbents for removal of heavy metal ions from solutions: a review. Materials 7(2):673–726
74. Sanchez C, Galo J, Ribot F, Grosso D (2003) Design of functional nano-structured materials through the use of controlled hybrid organic–inorganic interfaces. C R Chim 6(8–10):1131–1151
75. Schwarzenbach RP, Escher BI, Fenner K, Hofstetter TB, Johnson CA, Von Gunten U, Wehrli B (2006) The challenge of micropollutants in aquatic systems. Science 313(5790):1072–1077
76. Sharma P, Jindal R, Maiti M, Jana AK (2016) Novel organic–inorganic composite material as a cation exchanger from a triterpenoidal system of dammar gum: synthesis, characterization and application. Iran Polym J 25(8):671–685

77. Snyder A, Bo Z, Moon R, Rochet J-C, Stanciu L (2013) Reusable photocatalytic titanium dioxide–cellulose nanofiber films. J Colloid Interface Sci 399:92–98
78. Somasundaran P, Runkana V, Kapur PC (2005) Flocculation and dispersion of colloidal suspensions by polymers and surfactants: Experimental and modeling studies. Coagul Floccul 126:767–803
79. Souza VC, Quadri MGN (2013) Organic-inorganic hybrid membranes in separation processes: A 10-year review. Brazilian Journal of Chemical Engineering 30(4):683–700
80. Sun W, Long J, Xu Z, Masliyah JH (2008) Study of Al(OH)3-polyacrylamide-induced pelleting flocculation by single molecule force spectroscopy. Langmuir 24(24):14015–14021
81. Supply WJW, Programme SM, Organization WH (2015) Progress on sanitation and drinking water: 2015 update and MDG assessment. World Health Organization
82. Tetteh, E. K., & Rathilal, S. (2018) Evaluation of the coagulation floatation process for industrial mineral oil wastewater treatment using response surface methodology (RSM). Int J Environ Impacts 1(4):491–502
83. Thue PS, Sophia AC, Lima EC, Wamba AG, de Alencar WS, dos Reis GS, Rodembusch FS, Dias SL (2018) Synthesis and characterization of a novel organic-inorganic hybrid clay adsorbent for the removal of acid red 1 and acid green 25 from aqueous solutions. J Clean Prod 171:30–44
84. Tongwen X (2002) Electrodialysis processes with bipolar membranes (EDBM) in environmental protection—a review. Resour Conserv Recycl 37(1):1–22. https://doi.org/10.1016/S0921-3449(02)00032-0
85. Tzoupanos ND, Zouboulis AI (2011) Preparation, characterisation and application of novel composite coagulants for surface water treatment. Water Res 45(12):3614–3626
86. Ulbricht M (2006) Advanced functional polymer membranes. Polymer 47(7):2217–2262
87. Uzun I, Güzel F (2000) Adsorption of some heavy metal ions from aqueous solution by activated carbon and comparison of percent adsorption results of activated carbon with those of some other adsorbents. Turk J Chem 24(3):291–298
88. Vörösmarty CJ, McIntyre PB, Gessner MO, Dudgeon D, Prusevich A, Green P, Glidden S, Bunn SE, Sullivan CA, Liermann CR (2010) Global threats to human water security and river biodiversity. Nature 467(7315):555–561
89. Wang W, Ding Z, Cai M, Jian H, Zeng Z, Li F, Liu JP (2015) Synthesis and high-efficiency methylene blue adsorption of magnetic PAA/MnFe2O4 nanocomposites. Appl Surf Sci 346:348–353
90. Wang Y, Gao B, Yue Q, Wei J, Li Q (2008) The characterization and flocculation efficiency of composite flocculant iron salts–polydimethyldiallylammonium chloride. Chem Eng J 142(2):175–181
91. Wang Y, Shi R, Lin J, Zhu Y (2011) Enhancement of photocurrent and photocatalytic activity of ZnO hybridized with graphite-like C 3 N 4. Energy Environ Sci 4(8):2922–2929
92. Wang L, Wu X-L, Xu W-H, Huang X-J, Liu J-H, Xu A-W (2012) Stable organic–inorganic hybrid of polyaniline/α-zirconium phosphate for efficient removal of organic pollutants in water environment. ACS Appl Mater Interfaces 4(5):2686–2692
93. Wang Li, Zhang J, Wang A (2011) Fast removal of methylene blue from aqueous solution by adsorption onto chitosan-g-poly (acrylic acid)/attapulgite composite. Desalination 266(1–3):33–39
94. Wang Y, Gao BY, Yue QY, Wei JC, Zhou WZ (2006) Novel composite flocculent ployferric chloride-polydimethyldiallylammonium chloride (PFC-PDMDAAC): its characterization and flocculation efficiency. Water Pract Technol 1(3)
95. Weiss CK, Landfester K (2010) Miniemulsion polymerization as a means to encapsulate organic and inorganic materials. In: Hybrid latex particles, pp 185–236. Springer
96. Whittell GR, Hager MD, Schubert US, Manners I (2011) Functional soft materials from metallopolymers and metallosupramolecular polymers. Nat Mater 10(3):176–188
97. Wu J-B, Yi Y-L (2013) Removal of cadmium from aqueous solution by organic-inorganic hybrid sorbent combining sol-gel processing and imprinting technique. Korean J Chem Eng 30(5):1111–1118

98. Xu T (2005) Ion exchange membranes: State of their development and perspective. J Membr Sci 263(1–2):1–29
99. Yagci Y (2012) New photoinitiating systems designed for polymer/inorganic hybrid nanocoatings. J Coat Technol Res 9(2):125–134
100. Yan L, Li YS, Xiang CB (2005) Preparation of poly (vinylidene fluoride)(pvdf) ultrafiltration membrane modified by nano-sized alumina (Al2O3) and its antifouling research. Polym 46(18):7701–7706
101. Yang WY, Qian JW, Shen ZQ (2004) A novel flocculant of Al (OH) 3–polyacrylamide ionic hybrid. J Colloid Interface Sci 273(2):400–405
102. Yang Y, Wang P (2006) Preparation and characterizations of a new PS/TiO2 hybrid membranes by sol–gel process. Polym 47(8):2683–2688
103. Yu B, Guo S, He L, Bu W (2013) Synthesis and characterization of a luminescence metallosupramolecular hyperbranched polymer. Chem Commun 49(32):3333–3335
104. Zewail TM, El-Garf SAM (2010) Preparation of agriculture residue based adsorbents for heavy metal removal. Desalin Water Treat 22(1–3):363–370
105. Zhang K, Zha Y, Peng B, Chen Y, Tew GN (2013) Metallo-supramolecular cyclic polymers. J Am Chem Soc 135(43):15994–15997
106. Zhang Y, Zhao X., Li X., Liu C, Zhu L (2011) Algae-removal efficiency of AS/PDM used for the Taihu Lake Prechlorination Algae-Rich water in summer. J Chem Eng Chin Univ 2
107. Zhu T, Lin Y, Luo Y, Hu X, Lin W, Yu P, Huang C (2012) Preparation and characterization of TiO_2-regenerated cellulose inorganic–polymer hybrid membranes for dehydration of caprolactam. Carbohyd Polym 87(1):901–909
108. Zhu Y, Zhang X, Li R, Li Q (2014) Planar-defect-rich zinc oxide nanoparticles assembled on carbon nanotube films as ultraviolet emitters and photocatalysts. Sci Rep 4:4728
109. Zou J, Zhu H, Wang F, Sui H, Fan J (2011) Preparation of a new inorganic–organic composite flocculant used in solid–liquid separation for waste drilling fluid. Chem Eng J 171(1):350–356

Hydroxyapatite-Based Materials for Environmental Remediation

Abdallah Amedlous, Mohammed Majdoub, Othmane Amadine, Younes Essamlali, Karim Dânoun, and Mohamed Zahouily

Abstract Hydroxyapatite with a general chemical formula $Ca_{10}(PO_4)_6(OH)_2$ have experienced a growing interest in recent years as multifunctional biomaterials for a wide range of applications including dental implants, drug delivery vector, catalyst as well as various other fields for environmental remediation. Hydroxyapatite (HAp) based materials can be hugely useful in the field of pollution control owing to their attractive physicochemical properties, especially, high adsorption capacity, excellent chemical and thermal stability, ion-exchange capacity, acid–base properties, non-toxicity, and abundantly available in nature. This chapter provides on the one hand an overview about the synthetic methods for preparation of hydroxyapatite with description of textural and morphological proprieties. On the other hand, the different proprieties of HAp will also be discussed. Finally, the application of HAp in the field of environmental remediation including adsorption of heavy metals and organic molecules, catalysis, photocatalysis for organic pollutant degradation and membranes for oil/water separation will be described and recapped.

Keywords Hydroxyapatite · Environment remediation · Organic pollutant degradation · Photocatalysis · Adsorption · Oil/water separation

1 Introduction

Hydroxyapatite (HAp) is a calcium phosphate, whose general formula is $Ca_{10}(PO_4)_6(OH)_2$ with a stoichiometric Ca/P ratio of 1.67. Particularly, HAp's form is frequently hexagonal, with a $P6_3/m$ space group symmetry, similar to the mineral

A. Amedlous (✉) · M. Majdoub · M. Zahouily
Laboratory of Materials, Catalysis & Valorization of Natural Resources, Hassan II University, FST-Mohammedia, B.P. 146, 20650 Casablanca, Morocco
e-mail: abdallah.amedlous@ensicaen.fr

O. Amadine · Y. Essamlali · K. Dânoun · M. Zahouily
MASCIR Foundation, Rabat Design, Rue Mohamed El Jazouli, Madinat El Irfane 10100, Rabat, Morocco

E. Lichtfouse et al. (eds.), *Inorganic-Organic Composites for Water and Wastewater Treatment*, Environmental Footprints and Eco-design of Products and Processes,
https://doi.org/10.1007/978-981-16-5916-4_3

constituent of human teeth and bones [1]. The hydroxyapatite materials are characterized by their excellent physicochemical proprieties thanks to their, ion exchange ability, chemical and thermal stability, non-toxicity, biocompatibility, acid–base properties and high adsorption capacity. HAp materials are usually synthesized from calcium- and phosphorus-containing compounds via different methods including co-precipitation, sol–gel and hydrothermal. In addition, HAp can also be derived from natural sources, and waste by-products, which make them cost effective and environmentally friendly materials. It is owing to these properties that HAp is applied in numerous fields including dental implants [2], drug delivery vector [3], adsorption of organic and inorganic compounds [4–6], catalyst for various chemical transformations [7–9], photocatalysis [10–13] and as slow-release fertilizer for agricultural applications [14, 15]. In the field of water treatment, several researchers have reported the modification of HAp by other functional materials (e.g. organic and inorganic compounds, metals ion, carbon materials and biopolymers) to form novel composites. The obtained materials have additional advantages due to the synergistic properties between HAp and other functional materials, which make them as promising candidate for adsorption of various inorganic and organic compounds, oil/water separation, and catalyst/photocatalyst for the degradation of organic contaminant in aqueous medium.

The current chapter summarizes and overviews hydroxyapatite based materials and their environmental application. Herein, we are providing insights on the relationship between the structural, textural and morphological properties of HAp and the adopted synthetic approaches including co-precipitation, sol–gel, hydrothermal and microwave-assisted method. Furthermore, the environmental applications of hydroxyapatite based materials including adsorption, oil/water separation, catalysis and photocatalysis are discussed and recapped. From this chapter, we works towards furnishing the know-how in HAp chemical synthesis and modification approches for environment remediation purposes and stimulate the readership to seek novel HAp synthesis and modification strategies for new environment-related applications.

2 Common Routes for Synthesis of HAp

The preparation of hydroxyapatite via wet-chemical method concerns sol–gel, co-precipitation, hydrothermal and emulsion approach, which are considered as the more applicable routes due to their simplicity and reasonable cost. The most significant aspect of these approaches resides in their potential to control the crystallinity, porosity, particles size and shape as well as morphology. These methods use different types of solvents, temperatures, pressure and diverse chemical reagents and auxiliary additives. However, the above-mentioned process suffer from some limitations mostly related to the production, in some cases, of different calcium phosphates phases which give rise to the problem of purity. In the following section, we have highlighted four strategies to synthesis HAp using wet chemical method. In addition, the experimental procedure, particle size and textural properties are discussed and recapped.

2.1 Co-Precipitation

Currently, chemical precipitation is considered as the most common method for Hap preparation owing to the simplicity of the procedure. The experimental procedure consists of reaction between phosphorus and calcium ions under well controlled temperature and pH. Generally, Ca^{2+} and $(PO_4)^{3-}$ salts with a molar ratio of Ca/P = 1.67 are mixed in an aqueous solution, and then the mixture are precipitated by the addition of a base such as sodium hydroxide or ammonium hydroxide. It proceeds with several step: (1) precipitation (2) aging (3) separation (4) drying and (5) calcination. The chemical reactions of Hap precipitation is shown as follows:

$$5\,Ca(OH)_2 + 3H_3PO_4 \rightarrow Ca_5(PO_4)_3(OH) + 9H_2O \tag{1}$$

$$5\,Ca(NO_3)_2 + 3(NH_4)_2HPO_4 + NH_4OH \rightarrow Ca_5(PO_4)_3(OH) + 10\,NH_4NO_3 + 3\,H_2O \tag{2}$$

These two chemical reactions turned out to be especially popular due to their level of repeatability and relative ease. However, the synthesis of HAp using the first equation is mostly used due to formulation is simple and its only byproduct is water.

The degree of crystallinity, specific surface area, shape and size of the obtained HAp particles rest mainly on (1) the genre of reagents used (2) the pH of solution (3) the reaction temperature (4) the reaction time, (5) precipitation rate, (6) drying method (7) type of solvent and (8) calcination temperature [16, 17]. For example, Kumar and collaborators [18] have conducted a study on the effect of reaction temperature during precipitation on morphological changes. Co-precipitation technique was used to fabricate the Hap material employing calcium calcium hydroxide and *ortho*-phosphoric acid as sources of Ca and P, respectively, at distinct temperatures. As given in Fig. 1a, when the precipitation was conducted at 40 °C, the particle morphology is needle-like. However, during precipitation at 80 °C and 100 °C the

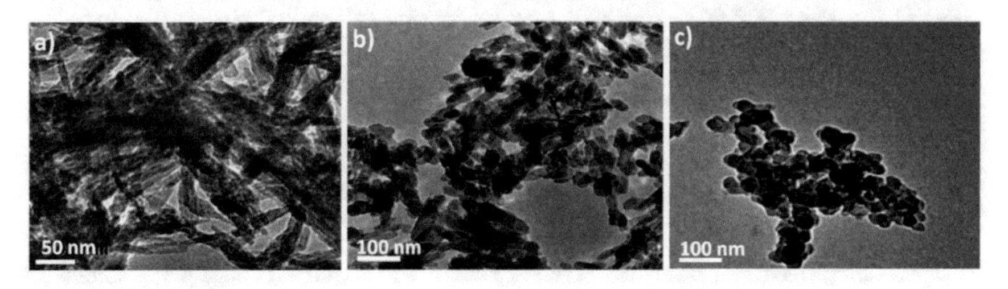

Fig. 1 Transmission electron microscopy (TEM) images of hydroxyapatite precipitated at distinct temperatures **a** 40 °C, **b** 80 °C, and **c** 100 °C, Reproduced with permission from [18]. Copyright 2004, American Chemical Society

particle change their morphology to pyramidal and spheroidal, respectively (Fig. 1b-c). Furthermore, Wang et al. [17] described the influence of different parameters on the morphology of HAp particles such as solvents, dispersants and drying method. The authors used diammonium phosphate and calcium nitrate as P and Ca sources for HAp synthesis using precipitation route in aqueous medium. They reported obtaining different morphologies by controlling the preparation conditions including needle-, rod-, sphere-, wire- and bamboo-leaf-like shapes.

2.2 *Hydrothermal*

The hydrothermal method also named solvothermal when the synthesis is performed in organic solutions is considered as one of the most method for preparation of pure and mono-dispersed homogeneous nanoparticles with high crystallinity. This technique is similar to chemical precipitation; however, the maturation step occurred in an autoclave, or in specific reactor, under pressure and temperature, which is generally above the boiling point of water.

Li et al. [19] reported the preparation of HAp nanorods via hydrothermal method. The particle size was controlled by different factors including the time of reaction and the type of alkaline solution. After 8 h of hydrothermal treatment, the obtained particles using NaOH solution supplying OH^-, appeared in the form of rods having a length in the range of 150–300 nm, as shown in Fig. 2a. However, belt-like particles with lengths ranging from 600–1500 nm were obtained after increasing of reaction time to 12 h, as presented in Fig. 2b. Contrarily, short nanorods of HAp (80–180 nm) were obtained when ammonia solution was used to adjust the pH of solution after 8 h of hydrothermal treatment (Fig. 2c). In addition, decreasing time of reaction

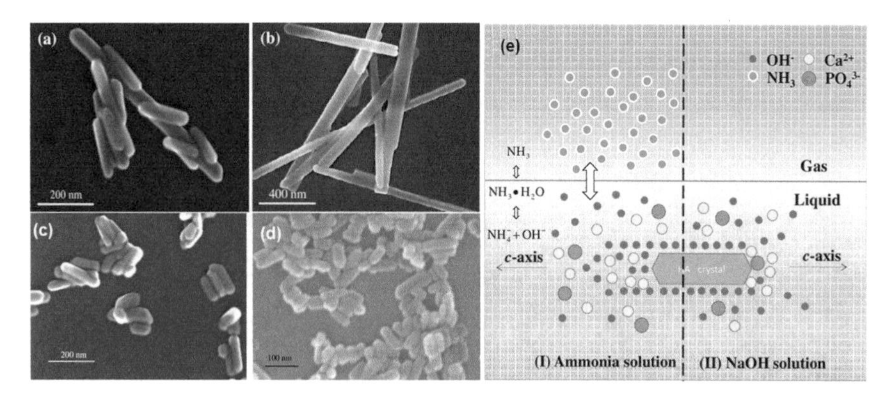

Fig. 2 Scanning electron microscopy (SEM) of HAP using NH_4OH to adjust pH of solution: **a** hydrothermal treatment of 8 h, **b** 12 h, SEM image of HAP using NaOH to adjust pH of solution: **c** hydrothermal treatment of 8 h, **d** 2 h, **e** schematic of mechanism of HAP particles. Reproduced with permission from [19]. Copyright 2017, American Chemical Society

to 2 h lead to formation of short nanorods of HAp (60–100 nm) (Fig. 2d). The mechanism of controlled of size was explained by idea that NH_4OH solution can provide more stable alkalinity in the hydrothermal treatment because is volatilized from the solution leading constant gas pressure. Therefore, OH^- was constantly consumed and supplied in the reaction, which made a significant effect on the growth of HAp crystal grain.

Qiao et al. [20] investigated the synthesis of hollow mesoporous carbonated HAp microsphere under hydrothermal conditions using hollow mesoporous $CaCO_3$ precursors. Firstly, the authors prepared hollow mesoporous $CaCO_3$ precursors using $CaCl_2$, Na_2CO_3 with different concentration sodium dodecyl sulfate (SDS) as a surfactant. The morphology of $CaCO_3$ was significantly affected by the concentration of SDS, as illustrated in Fig. 3a–d. The increasing of SDS concentration from 1 mg mL^{-1} to 10 lead to the increasing of size of microspheres from around 2 mm to more than 5 mm. In addition, the hollow space in the spheres was also expanded. The preparation of HAp microspheres with different $CaCO_3$ is presented in Fig. 3i. The obtained hollow $CaCO_3$ precursors were mixed with Na_2HPO_4, and the pH of mixture was adjusted with 0.2 M of NH_4OH solution until pH 11. Finally, the mixture was transferred into a Teflon-lined stainless steel autoclave for a hydrothermal reaction at 180 °C for 24 h. The morphology of as-prepared HAp samples showed microspheres-type morphology with a hollow space in the core of microspheres (as observed in Fig. 3g–h). Furthermore, the mesoporous structures were confirmed by nitrogen sorption method. The pore structures in the microspheres increased with increase of concentration of SDS. The pore size distribution obtained using BJH method was in the range of 3–100 nm. The increasing of SDS concentration from 1 mg/mL to 10 mg/mL lead in increasing of the surface area from 17 m^2/g to 25 m^2/g, respectively. The different morphology of HAp microspheres was tested in the drug release properties in a pH dependent manner.

2.3 *Sol–Gel Method*

Among the alternative methods, sol–gel involve atomic level molecular mixing, providing control of composition and chemical homogeneity [21]. This technique is a representative of wet chemistry to prepare variety of materials with high purity using low synthesis temperature [22–24]. The general route to prepare HAp via sol–gel necessitates an exact molar ratio of 1.67 between calcium and phosphorous sources. Various starting materials may be used in the synthesis of HAp via sol–gel including calcium dioxide (CaO), calcium nitrate ($Ca(NO_3)_2$), calcium diethoxide (Ca(OEt)) or calcium acetate ($Ca(C_2H_3O_2)_2$) as calcium precursors and triethyl phosphite ($(C_2H_5O)_3P$), triethyl phosphate ($PO(OEt)_3$) phenyldichlorophosphite (C_6H_5PCl) or phosphonoacetic acid (HOOCCHPO(OH)) as phosphorous precursors [24–27].

For instance, Liu et al. [24] used tiethyl phosphite to synthesis the phosphorus sol by hydrolyzing the phosphite in a mixture of ether/water or ethanol/small quantity of water, followed by addition in a fast manner of calcium precursor. The results showed

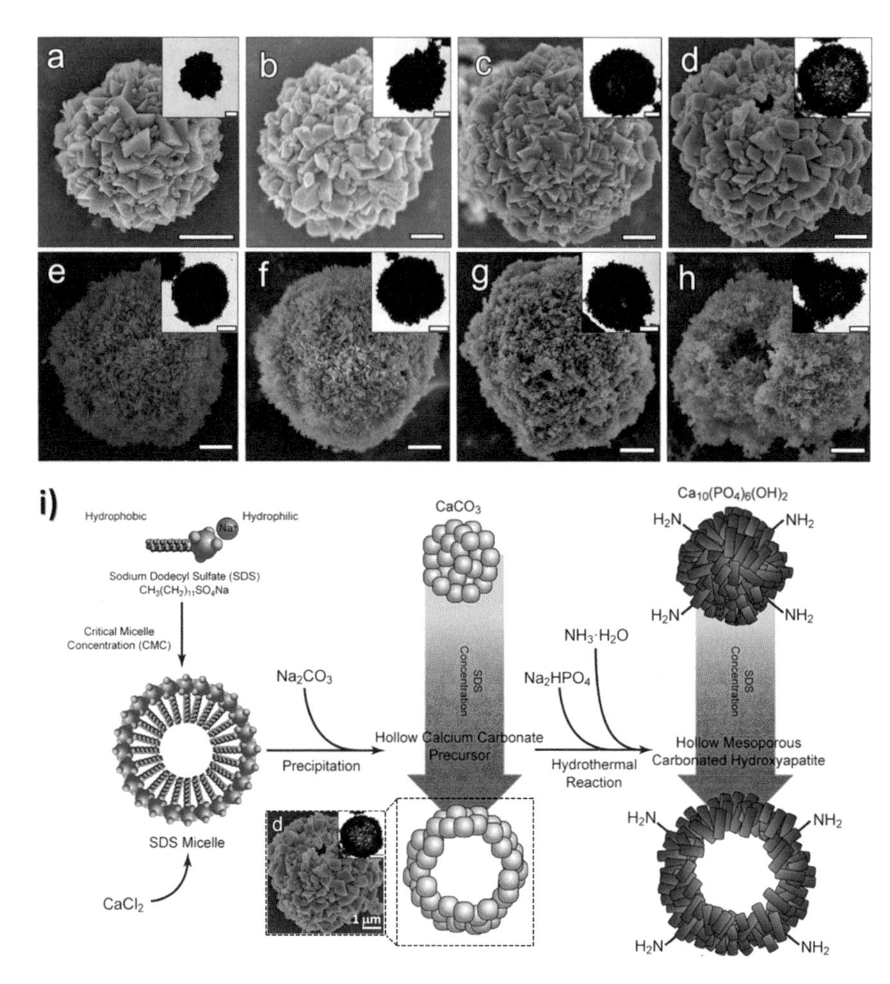

Fig. 3 SEM micrographs with inserted TEM images for $CaCO_3$ precursors of **a** A-CHAM (SDS = 1 mg/L), **b** B-CHAM (SDS = 2.33 mg/mL), **c** C-CHAM (SDS = 4.66 mg/mL) and **d** D-CHAM (SDS = 10 mg/mL); SEM micrographs with inserted TEM images showing the morphology of **e** A-CHAM, **f** B-CHAM, **g** C-CHAM and **h** D-CHAM (scale bar = 1 mm), **i** Schematic diagram showing the synthesis of hollow mesoporous carbonated HAp microsphere (CHAMs). Reproduced with permission from [20]. Copyright 2017, Royal Society of Chemistry

that pure HAp phase was formed at low temperature of 350 °C. The ethanol-based method appears to provide thermally stable HAp. However, calcium-deficient apatite was formed using water. The authors also reported the preparation of porous HAp coating by deposition of gel onto a Ti metal substrate. Conversely, Brendel et al. [28] prepared HAp using calcium nitrate and phenyldichlorophosphine as initial compounds. The resulting HAp product obtained at 400 °C show poor degree of crystallinity and low purity. Nonetheless, the increasing of calcination temperature up to 1000 °C can promote the formation of pure and crystallized HAp phase.

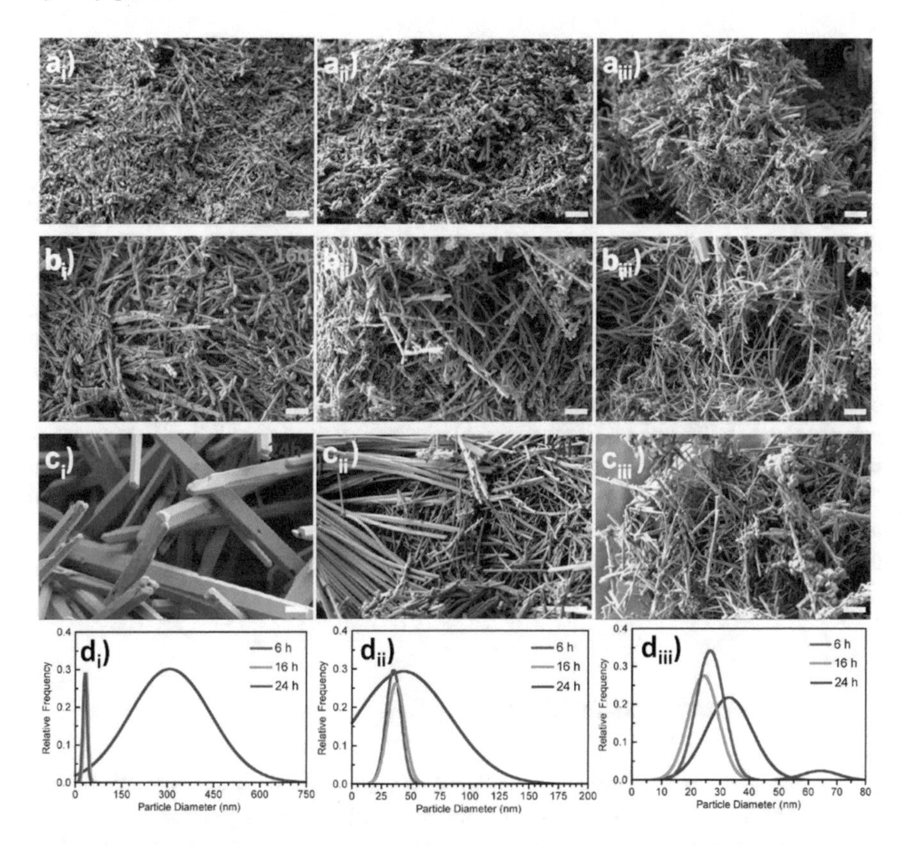

Fig. 4 HAp particles' SEM micrographs and particle diameter distribution. Hydrothermal reaction conducted at pH = 5.20 **ai** 6 h, **bi** 16, **ci** 24 h and **di** particle diameter distributions; Hydrothermal reaction conducted at pH = 7.50 **aii** 6 h, **bii** 16, **cii** 24 h and **dii** particle diameter distributions; Hydrothermal reaction conducted at pH = 13.70 **aiii** 6 h, **biii** 16, **ciii** 24 h and **diii** particle diameter distributions. Reproduced with permission from [1]. Copyright 2012, American Chemical Society

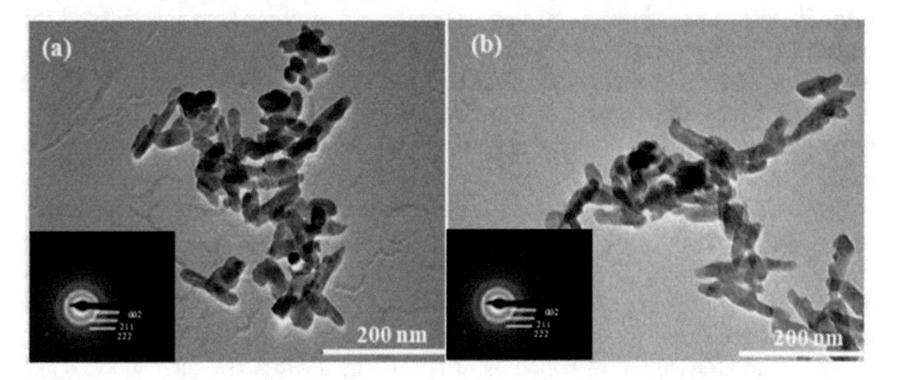

Fig. 5 TEM images of HAp particles prepared by microwave assisted method using SLES and (b) LABS surfactants. Reproduced with permission from [29]. Copyright 2014, Royal Society of Chemistry

In another study, Costa et al. [1] developed novel method based on combining sol–gel and hydrothermal process for preparation of micro and nanoscale HAp particles with different morphologies. The gel was prepared using similar method to that developed by Liu et al. [24]. The obtained dried amorphous calcium phosphate (ACP) gel was suspended in different initial pH of aqueous solution (5.2, 7.5 or 13.7). Finally, for each aqueous solution with different pH were subjected to hydrothermal treatment for a period of 6, 16, or 24 h. The results indicate the formation of small nanorods with diameters of 30 nm at pH of 5.20 and hydrothermal treatment of 6 and 16 h (Fig. 4ai–biai–bi). However, increasing time of reaction to 24 h resulted in increasing of particles diameter to 300 nm, as shown in (Fig. 4ci). When the suspension's initial pH increased to 7.50, the particles' diameter exhibited a decrease compared to particles synthesized under pH of 5.2. Short nanorods-like morphology was obtained by Six hours of hydrothermal treatment (Fig. 4aii, diiaii, dii). Nanowire-like morphology was produced via the extention of hydrothermal treatment to hours (Fig. 4bii and diibii and dii). In contrast, extending time of reaction to 24 h resulted in an increase in particle diameter (Fig. 4dii). However, the particle diameter was found to decrease, when initial pH of suspension was increased to 13.70 (ranging from 15 to 100 nm) (Fig. 4aiii–diiiaiii–diii).

2.4 Microwave-Assisted Method

Microwave-assisted methods is considered as an emerging new approach for the synthesis of homogenous size, porosity and morphology of HAp nanoparticles. This process can be described as chemical precipitation occurring with microwave heating.

Amer et al. [29] studied the fabrication of mesoporous HAp using two anionic surfactants (linear alkylbenzenesulfonate (LABS) and sodium lauryl ether sulfate (SLES)) via microwave-induced reaction.

The TEM images of prepared HAp using SLES (Fig. 5a) and LABS (Fig. 5b)surfactants showed rod-like nanoparticles. Authors also investigated the outcome of using LABS and SLES as template for HAp nanoparticles' synthesis, the surface area was found to increase from 30 $m^2\ g^{-1}$ for pristine HAp to 48.26 and 60.6 $m^2\ g^{-1}$ using SLES and LABS surfactants, respectively. The pore formed after removal of surfactants was calculated using BJH method was around 35 nm, which confirm the formation of mesoporous HAp materials.

In another research study, which used microwave-induced hydrothermal reaction to fabricate hierarchically nanostructured porous hollow spheres of HAp [30]. The authors used organic phosphorus source and calcium chloride. The structural characterization of materials using XRD showed that all the peaks were characteristics of single-phase HAp. The surface area was 87 m^2/g and pore diameter of 20.6 nm, which is characteristic of mesoporous material. The authors also studied the effect of using organic phosphorus source by SEM and TEM analysis. By using Na_2HPO_4 as an inorganic phosphorus source, a mixture of nanosheets nanorods were found as displayed in Fig. 6a–b. However, hierarchically assembled into nanostructured

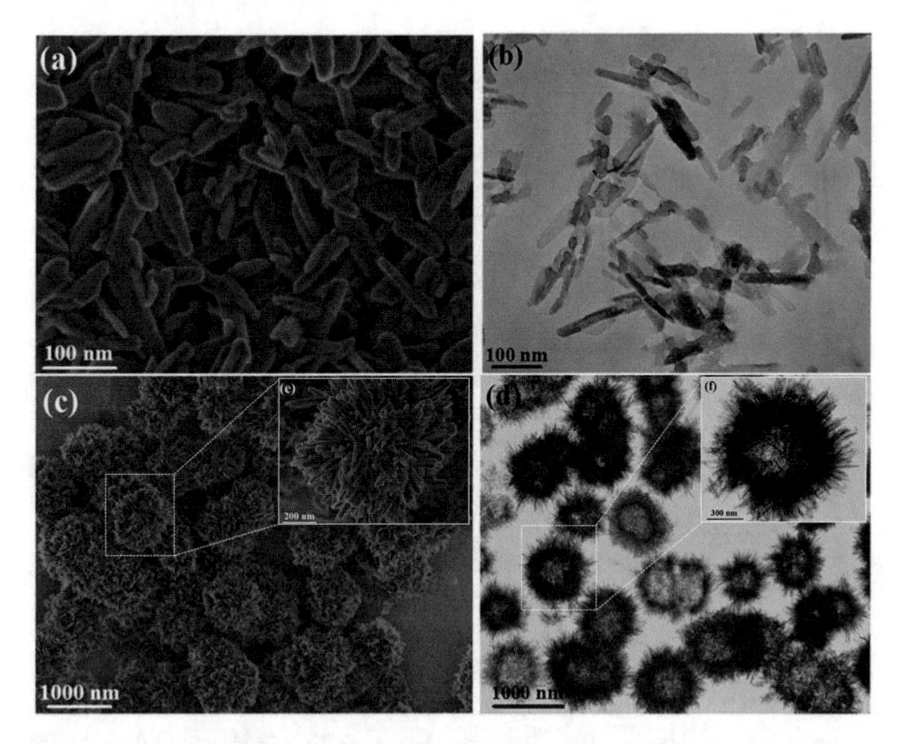

Fig. 6 HAp particles' SEM and TEM micrographs obtained by microwave-induced hydrothermal technique. Reproduced with permission from [30]. Copyright 2013 Wiley and Sons

porous spheres morphology was obtained using another P source which was creatine phosphate, as shown in Fig. 6c a–d. These reported findings highlights that the P source takes part in the final HAp's morphology.

3 Proprieties

3.1 Crystalline Structure

Hydroxyapatite is considered as a calcium phosphate having a formula of $Ca_{10}(PO_4)_6(OH)_2$, which has a stoichiometric Ca/P ratio of 1.67 and crystallizes in the hexagonal phase with $P6_3/m$ space group and lattice parameters a = b = 9.432 Å, and c = 6.881 Å (JCPDS No. 09–432) [31]. The unit cell of HAp is composed by calcuim (Ca^{2+}) and phosphates (PO_4^{3-}) and can be represented by $Ca(I)_4$ $Ca(II)_6(PO_4)_6(OH)_2$. In fact, as represented in Fig. 7, the Ca (I) sites are hexa-coordinated with oxygen coming exclusively from the phosphate groups. Conversely, the Ca(II) sites are coordinated with seven oxygen atoms, one of which is that of a

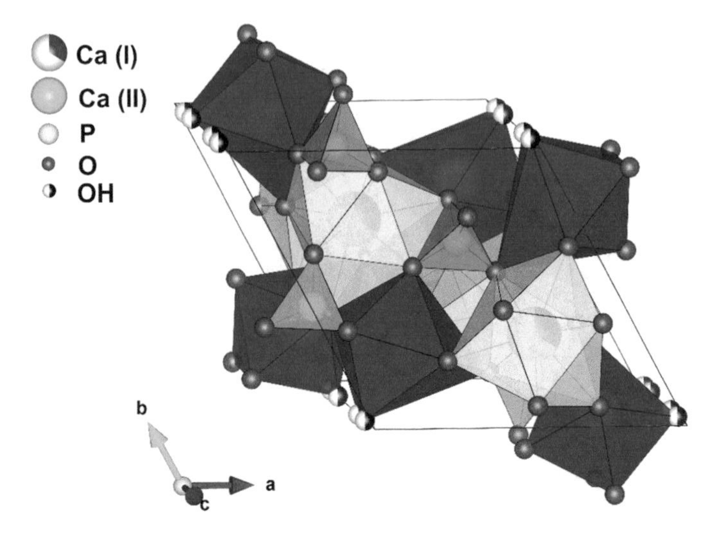

Fig. 7 Projection showing the sequence of octahedral: [$Ca(I)O_6$] and [$Ca(II)O_6$] and tetrahedral [PO_4] in the HAp structure drawn using VESTA software

hydroxyl group. In addition, HAp can also exist in another form, that is the monoclinic phase with $P2_1/b$ space group and lattice parameters a = 9.4214 Å, b = 2a Å and c = 6.8814 Å (JCPDS No. 76–0694) [32]. This form is known to be the most ordered and thermodynamically stable even at room temperature. The orientations of OH groups are the main disctinction between HAp's hexagonal and monoclinic phases. In the case of HAp with a hexagonal structure: the OH groups occur in columns on the screw axes, and the adjacent OH point in opposite directions. On the other hand, in monoclinic phase, all of the OH in a given column are pointed in the same direction, and the direction reverses in the next column [32].

3.2 *Substitution Properties*

One important propriety of the apatite lattice is its flexibility, which give it great adaptability in terms of substitution ability both in the calcium site and in the phosphorus or hydroxyl site.

3.2.1 Cationic Substitution

In the apatite structure, calcium sites with ionic radius of 0.99 Å can be doped or substituted by various metals whose size and charge may be different from that of Ca^{2+}. For example, (i) monovalent cation (Na^+, K^+, Ag^+ and Cs^+) (ii) divalent cation (Ni^{2+}, Co^{2+}, Cu^{2+}, Sr^{2+} and Cd^{2+}) (iii) trivalent cation (Fe^{3+} and La^{3+}) (iv) tetravalent

(Ti^{4+} and Zr^{4+}), and (v) pentavalent cation (Ta^{5+}, V^{5+} and Nb^{5+}) or vacancies [33–36]. However, not all cations can be substituted with the same facility and quantity, which is depended on the ionic radius and charge of cation [37–39]. It is acknowledged in the state of the art that cations with a radius larger than Ca^{2+} such as Sr^{2+}, Ba^{2+} and Pb^{2+} prefer to be incorporated in Ca(II) site. In contrast, cations with small radius including Zn^{2+}, Mg^{2+} prefer to occupy Ca(I) site [40]. The cationic substitution in hydroxyapatite can be induce changes in characteristics such as lattice parameters, crystallinity, particles size, thermal stability, textural and acid–base properties [41, 42]. Consequently, this makes it possible to consider materials for wide range of applications in various fields.

3.2.2 Anionic Substitution

The HAp material can also be substituted in anionic site with the total or partial substitution. For example, the hydroxyl site (OH^-) can be replaced by monovalent ions such as F^- and Cl^- without loss of charge. The carbonate ions ($CO_3{}^{2-}$) can be inserted into apatite framework in two types of anion positions OH- and $PO_4{}^{3-}$ i.e., the substitution in OH^- position known as A-type and in $PO_4{}^{3-}$ position as B-type carbonated apatites [43]. The B-type substitution is most preferred when the A/B ratio is between 0.7–0.9 [44].

Divalent, trivalent or tetravalent groups such as $HPO_4{}^{2-}$, $CO_3{}^{2-}$, $SO_4{}^{2-}$, $VO_4{}^{3-}$, $AsO_4{}^{3-}$, $MnO_4{}^{3-}$ and $SiO_4{}^{4-}$ etc. can also substitute in $PO_4{}^{3-}$ [45, 46]. The replacement of phosphate ions by other anions with different charges can cause imbalance charge, which must be accompanied by charge compensation. Indeed, the charge equilibrium is achieved by the formation of vacancies in both hydroxyl and calcium sites. In contrast, the existence of apatite with the presence of vacancies in $PO_4{}^{3-}$ sites has never been reported in the literature [47]. These phosphates are the largest groups of apatite structure and appear to be necessary for its stability. As example the illustration of substitution of $PO_4{}^{3-}$ groups by $HPO_4{}^{2-}$ and $SiO_4{}^{2-}$ are given below:

- Substitution of $PO_4{}^{3-}$ by $HPO_4{}^{2-}$ leads to the formation of vacancies in the Ca^{2+} and OH^- sites. The formula of this apatite is as follows: $Ca_{10\text{-}x}0_x(PO_4)_{6\text{-}x}(HPO_4)_x(OH)_{2\text{-}x}0_x$.
- Substitution of $PO_4{}^{3-}$ by $SiO_4{}^{4-}$ is accompanied by decreased in the hydroxyl content to obtain HAp with formula $Ca_{10}(PO_4)_{6\text{-}x}(SiO_4)_x(OH)_{2\text{-}x}0_x$.

The substitution of $PO_4{}^{3-}$ groups by other ions offer the possibility to modify the chemical properties of HAp, for example, the substitution by carbonate modify the acid–base proprieties and the substitution with vanadate can improve redox properties [48, 49].

As conclusion, the ionic substitutions of HAp can cause the modifications in lattice parameters, thermal stability, decrease in crystallinity and particle size, which can induce modifications of the surface properties.

3.3 Thermal Stability

The high thermal stability of materials is becoming an important property for several applications. Hydroxyapatite exhibited a good thermal stability, which is depends on several parameters including stoichiometry, substitution and synthesis conditions [50, 51]. From the data shown in Fig. 8, the precipitated HAp exhibits outstanding thermal stability (up to 800 °C in air). It has been reported that HAp material exhibit two to four stages of weight loss [52]. The first one in the temperature range of 25–200 °C assigned to the removal of physically adsorbed water without any effect on the lattice parameters. The second weight loss at temperature between 200–400 °C, attributed to the removal of lattice water, which causes a contraction in the a-lattice dimension during heating. At high temperature, the sintering of HAp will lead to a partial loss of OH, this process namely dehydroxylation which occurs at temperatures at about 900 °C in air atmosphere according to the following reaction (3) [53–55]:

$$Ca_{10}(PO_4)_6(OH)_2 \rightarrow Ca_{10}(PO_4)_6(OH)_{2-2x}O_x0_x + xH_2O\uparrow \tag{3}$$

where $0\times$ is a hydrogen vacancy.

The hydroxyl ion deficient product is a metastable crystal phase called oxyhydroxyapatite (OHAp) with the formula $Ca_{10}(PO_4)_6(OH)_{2\text{-}2x}O_x0_x$ with 0 vacancy and $x < 1$.

At very high temperatures, two OH combine to form one molecule of water, leaving a peroxy ion (O^{2-}) in the lattice and OHAp begins to decompose into other phases, following reaction (4) and (5) [56–58]:

$$Ca_{10}(PO_4)_6O \rightarrow 3\beta\text{-}Ca_3(PO_4)_2 + CaO + H_2O \tag{4}$$

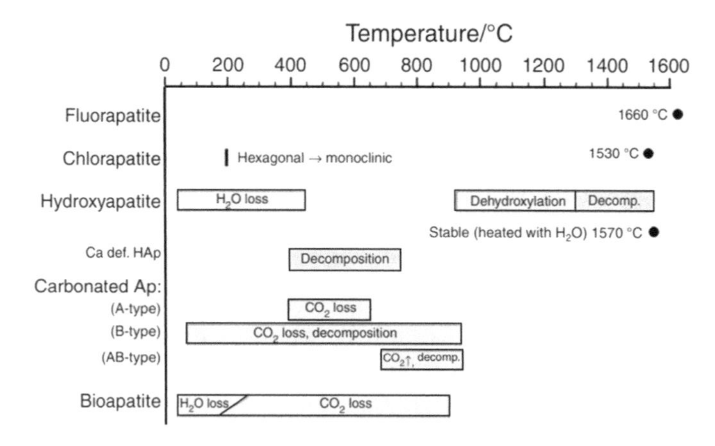

Fig. 8 Apatites thermal stages including FAp, ClAp, HAp, CO_3Ap, BioAp, displaying a loss of structural groups, phase change (ClAp), decomposition or melting (indicated by a black circle). Reproduced with permission from [52]. Copyright 2012 Springer Nature

$$Ca_{10}(PO_4)_6O \rightarrow 2\alpha\text{-}Ca_3(PO_4)_2 + Ca_4P_2O_9 + H_2O \tag{5}$$

3.4 PH Stability

The stability of HAp in a wide range of pH solution is another very important property. Kamieniak et al. [59] investigated the effect of the pH solution in the structural, textural and morphological proprieties of HAp. First, the authors confirmed that no morphological changes to the HAp in the acidic or alkaline conditions. In addition, XRD and FTIR spectroscopy show that hexagonally HAp structure was preserved by an exposure to aqueous buffer solution over the pH range of 2–12. However, the surface area of materials was strongly affected. The surface area increases in acidic media and decreases in basic media. The decrease in surface area at pH $\geq$ 8, can be explained by higher concentrations of OH^- ions in the surface of HAp, which may be cause some agglomeration of particles during the drying step, resulting in reduction of surface area. On the other hand, the solubility of HAp was studied by Bell et al. [60], the result indicate that solid HAp was insoluble over the pH range 4.56–9.67.

3.5 Acid–Base Propriety

Although HAp has been frequently used as catalyst or a catalyst support material for active phases for different acid–base reactions. This material exhibit both acid and base sites, depending on the conditions in which it is synthesized. The Ca/P atomic ratio of HAp play important role in acid–base adjustability in order to obtain the desired function. Several research groups have shown that HAp with Ca/P < 1.67 acts as acid catalyst, while Ca/P $\geq$ 1.67 acts as basic catalyst [61].

Hence, different studies have been carried out to identify the basic and acidic sites of HAp that are responsible to its chemical reactivity. Diallo-Garcia et al. [62] used diffuse reflectance infrared transform spectroscopy (DRIFTS) analysis to identify the basic Lewis and Brønsted sites of stoichiometric HAp using adsorption–desorption processes of Lewis acid (CO_2) and Brønsted acid (acetylene C_2H_2) molecules.

The results showed that hydrogenocarbonates ($HCO_3{}^-$) and surface carbonates ($CO_3{}^{2-}$) are formed during the adsorption of CO_2 (Fig. 9), which reveals the presence of basic Lewis sites such as OH^- and O^{2-} of the $PO_4{}^3$ groups, respectively (reaction 6 and 7).

$$CO_2 + OH^- \rightarrow HCO_3^- \tag{6}$$

$$CO_2 + O^{2-} \rightarrow CO_3^{2-} \tag{7}$$

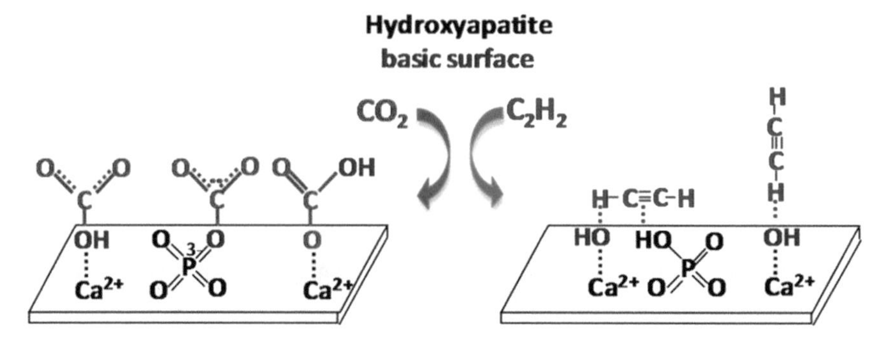

Fig. 9 Adsorption modes of CO_2 and acetylene on the surface of hydroxyapatite, Reproduced with permission from [62]. Copyright 2014 American Chemical Society

As for the acetylene probe, which was used to study Brønsted basicity, three non-dissociative adsorption modes of acetylene on the HAp surface were observed:

- π-type interaction with acidic P–OH
- An interaction with an acid–base pair (POH–OH),
- An interaction of type σ with OH^-.

On the other hand, the acidic sites are likely to be involved, two types can be considered, Lewis acid sites, which are calcium ions (Ca^{2+}), and Brønsted acid sites represented by P–OH.

4 Environmental Applications of Hydroxyapatite-Based Materials

4.1 Molecule Adsorption and Separation for Wastewater Remediation

4.1.1 Adsorption of Metals

Owing to the surge in industrial and mining activities, water contamination occurring by heavy metals such as copper (Cu^{2+}), lead (Pb^{2+}), mercury (Hg^{2+}), cadmium (Cd^{2+}), zinc (Zn^{2+}), etc., have caused a rise in human exposure to these toxic heavy metals. These pollutants, even at low concentrations, are highly dangerous for human beings, aquatic environment, and ecosystem as a whole [63]. Among the large variety of adsorbents materials, HAp based materials have been considerably investigated for removal of heavy metals ions from contaminated water via adsorption process. Owing of advantages of HAp including non-toxicity, high adsorption capacity, thermal stability and high stability over large pH range was expected as promising candidate for environmental application. The adsorption capacity of HAp was affected by several parameters including Ca/P stoichiometry, morphology, compositional, textural and structural properties [64–67]. In this context, Campisi et al. [67] prepared different HAp materials by precipitation method in aqueous solution with different Ca/P atomic ratio of 1.67 and 0.9, namely, stoichiometric HAp (S-HAP) and Ca-deficient HAp (d-HAP) and stoichiometric HAp precipitated from hydroalcoholic solution (a-HAP). The results indicated that all prepared HAp materials "maintained their high ability towards the Pb^{2+} capture, even in binary metal solutions containing both Pb^{2+} and Cu^{2+}. However, d-HAp is selected as the best adsorbent for elimination of Cu^{2+} in water with removal efficiency of 93%, compared to the s-HAP and a-HAP (74% and 80%, respectively). In binary system (presence of both Cu^{2+} and Pb^{2+}), d-HAP exhibited its high ability for adsorption of Cu^{2+}, even in the presence of a double concentration of Pb^{2+} compared to its own. The structural, textural results for adsorbents before and after adsorption tests suggest the two different mechanisms of adsorption. The removal of Pb^{2+} occurs through a dissolution–precipitation mechanism with formation of $Pb_{10}(PO_4)_6(OH)_2$ phase, whereas Cu^{2+} was immobilized by surface complexations (Fig. 10a) involving Ca^{2+} species with carbonate and hydroxyl groups of HAP.

In another study, Guan et al. [68] suggested that the adsorption mechanism of Pb^{2+} from aqueous solution using HAp involves not only precipitation and surface adsorption, but also incorporation, which depends on initial Pb^{2+} concentration. As shown in Fig. 10b, the authors suggested the presence of three mechanisms for lead uptake. At low concentrations of Pb^{2+} (e.g., 0.1 mM), the lead was totally adsorbed by surface functional groups of HAp. At medium concentration (0.5–5.7 mM), the dissolution of HAp occurred which allows to generate hydroxyls ions to increase pH of solution, thus generating a new metal phase precipitation ($Pb_5(PO_4)_3OH$, hydroxypyromorphite) via either OH^- or PO_4^{3-}. However, at high Pb^{2+} concentrations ($\geq$ 6.6 mM), ions exchange (incorporation of Pb^{2+} in the lattice) together with adsorption contributed to the overall uptake mechanisms.

Several other studies have tried to introduce new functional moieties HAP surface the enhancement of its adsorption capacity toward heavy metal ions [69–73]. For example, Wang et al. [71] developed a new strategy to regulate the pore structure and surface characteristic of esterified HAP (n-EHAP) nanocrystals. The authors prepared n-HAp and esterified HAP via conventional precipitation method using $Ca(NO_3)_2$, $4H_2O$, $(NH_4)_2HPO_4$, and 2-bromo-2- methylpropionate. The esterified HAP exhibited the high adsorption capacity of 2397 mg/g for removal of Pb^{2+} from water compared to pristine n-HAp (284 mg/g).

In another study, Oulguidoum et al. [74] developed a potential adsorbent involving the functionalization of HAp by sulfonate molecules. The modification of materials by sulfonic groups is well known as a powerful strategy to produce promising adsorbents for metals uptake. The authors of this study prepared sulfonated HAp material from natural phosphate via dissolution/precipitation approach. After dissolution of natural phosphate by nitric acid solution, an aqueous solution of sodium benzene-1,3-disulphonate (BDS) was added to the filtrate containing the calcium and phosphorus precursors and the pH of mixture was adjusted to 10 by ammonia solution. The sulfonated HAp exhibited adsorption capacities (qe) of 4.21 mmol g^{-1},

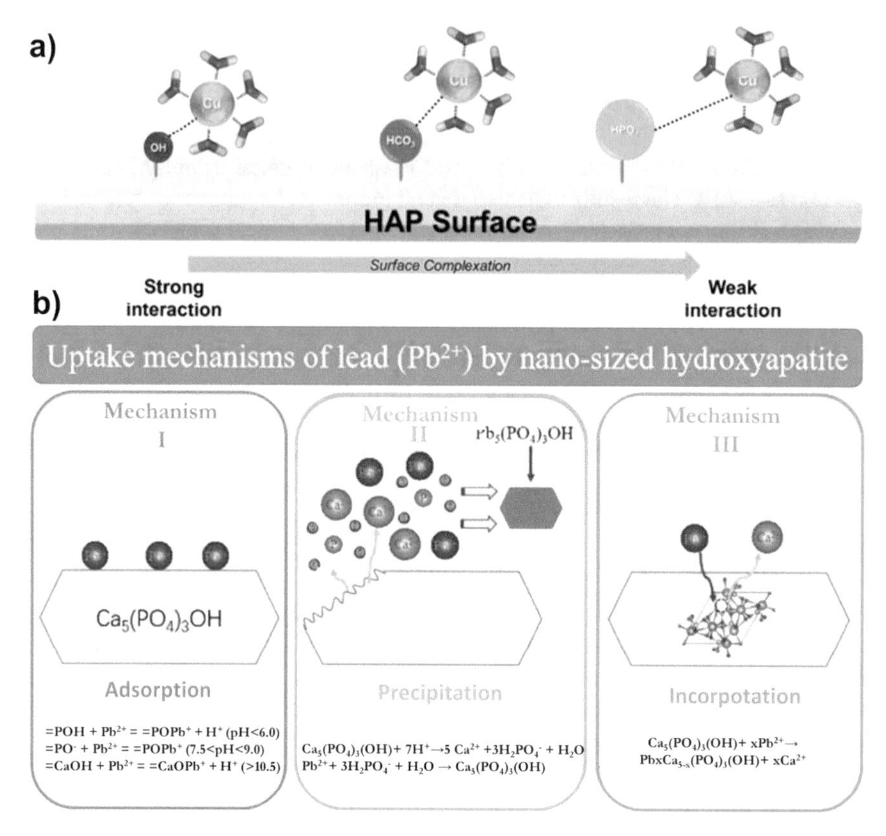

Fig. 10 **a** HAp surface anions' role on Cu^{2+} uptake through surface complexation mechanism, Reproduced with permission from [67]. Copyright 2018 Royal society of chemistry, **b** uptake mechanism of lead (Pb^{2+}) by nano-sized HAp, Reproduced with permission from [68]. Copyright 2018 American Chemical Society

3.15 mmol g^{-} and 2.45 mmol g^{-1} towards Zn^{2+}, Cd^{2+} and Pb^{2+}, respectively. Additionally, the sulfonated HAp exhibited antibacterial activities against E. coli and P. aeruginosa bacteria.

Considering that integration of phosphate ($PO_4{}^{3-}$) functional groups plays a key role in chelation of metals ions, Fang et al. [70] reported the preparation of zoledronate functionalized hydroxyapatite (zole-HAP) for heavy metals removal. In this work, authors prepared (zole-HAP) via in situ precipitation (Fig. 11a). First, certain amount of zole and $(NH_4)_2HPO_4$ was added into distilled water, the mixture was added drop by drop into $CaCl_2$ solution, the pH of mixture was adjusted to 10 by dropping NH_4OH. The presence of zoledronate in HAp matrix was confirmed by the FTIR spectroscopy, which proves the presence of C=C, C=N and P–OH groups. In addition, the cristallinity of HAp materials decreased as the added amount of zoledronate increase. This result can be explained by the interaction of bisphosphonate anion with calcium atoms through a bidentate chelation of deprotonated

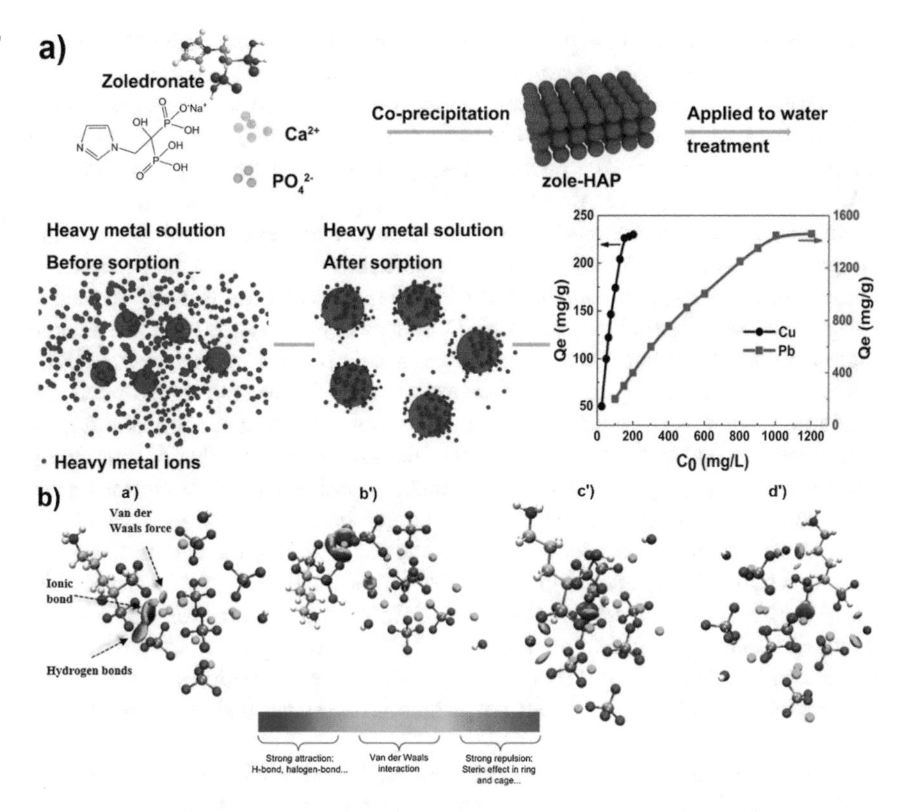

Fig. 11 **a** Preparation of zoledronate HAp and its application in water treatment, Reproduced with permission from [70]. Copyright 2020 Elsevier. **b** The isosurface of δg inter (isosurface value = 0.01 a.u.) **a** AL-HAP, **b**AL-HAP-Pb, **c**AL-HAP-Cu, **d**AL-HAP-Cd (Pb-gray, Cu-ochre, Cd-purple, Ca-green, C-cyan, P-tan, H-white, O-red, N-blue), Reproduced with permission from [75]. Copyright 2020 Elsevier

oxygen atoms of zoledronate, or hydroxyl group (OH) of bisphosphonate to partly form calcium zoledronate. The 10%zole-HAP material exhibited excellent adsorption capacity reached 1460.14 mg/g for Pb^{2+} and 226.33 mg/g for Cu^{2+}, respectively. The authors of this work suggested that adsorption mechanism of Cu^{2+} involves surface complexation and ion exchange, while dissolution–precipitation and surface complexation may contribute more in the adsorption of Pb^{2+}.

Using the same protocol, Ma et al. [75] synthesized alendronate (AL) molecules doped HAp. The Al-HAp composite doped with 10% of alendronate exhibited maximum adsorption capacity for Cu^{2+}, Cd^{2+}, and Pb^{2+} of 226.6 469, and 1431.8 mg/g, respectively. The adsorption mechanism was studied from the two aspects including experiments and quantum chemical calculation based on density functional theory (DFT). First, the authors of this work proposed mechanism of adsorption of three heavy metal ions by measuring initial and final pH of solution as well as concentration of Ca^{2+} ions in solution after completing the experiment. The results show increase on final pH solution when pH of initial was below five, which were mainly related to the release of OH^- and Ca^{2+}. However, when initial pH of solution was great than five, the final pH decreases to some extent after adsorption, which indicate that adsorption mechanism of Me^{2+} (Pb^{2+}, Cu^{2+} and Cd^{2+}) involve surface complexation. Therefore, the possible adsorption reaction formula is as follows: $HAP–OH_{(s)} + Me^{2+}{}_{(aq)} = HAP–O–Me^{2+}{}_{(s)} + H^+{}_{(aq)}$. This result was also confirmed by comparing XPS spectrums of adsorbent before and after experiment. The binding energy of O 1 s of Al-HAp after adsorption shifted, which is due to the chelation of a large number of oxygen-containing functional groups on the adsorbent surface to heavy metal ions. In addition, using XRD no new crystal phases have been found after adsorbing Cu^{2+} and Cd^{2+} ions. However, new phase $Ca_{2.5}Pb_{7.5}(PO_4)_6(OH)_2$ and $Pb_{10}(PO_4)_6(OH)_2$ appeared after adsorption of Pb^{2+}. The results confirmed that in addition to the surface complexation, the dissolution–precipitation occurred in lead adsorption. Using the DFT calculations, the authors confirm that AL has strong affinity with HAp (Fig. 11b, a′). The phosphonic acid groups of AL formed ionic bond with Ca^{2+} of HAP, and there were hydrogen bonds and van der Waals interactions with the phosphate group of HAP. After adsorption of metal ions, there was a strong a strong interaction among Pb^{2+}, $PO_4{}^{3-}$ of HAP, $HPO_3{}^-$ of AL and Ca^{2+} (Fig. 11b, b′). Similar situation in the case of Cu^{2+} but the interaction strength was obviously weaker than that of the Pb^{2+} (Fig. 11b, c′). Among them, Cd^{2+} interacted weakest with AL-HAP (Fig. 11b, d′). As conclusion, the interaction strength of AL-HAP metal ions is $Pb^{2+} > Cu^{2+} > Cd^{2+}$, which is correlated with the experimental results.

On the other hand, various research and studies showed that polysaccharide molecules can be used as promising candidate for the removal of a number of heavy metals because of its chelating ability towards metal ions [76–80]. In this regard, Manatunga and co-workers [81] designed a facile in-situ synthetic approach to synthesize HAp nanoparticles polymer nanocomposites using chitosan (CTS@HAP) and carboxymethyl cellulose (CMC@HAP). The results of this study demonstrated that these nanocomposites exhibited high adsorption capacity for Pb^{2+} uptake. The maximum adsorption capacity was found to be 625 mg/g (achieved in 3 min) and 909 mg/g (achieved in 30 s) using CMC@HAP and CTS@HAP, respectively.

In another paper, Saber-Samandari et al. [80] reported the synthesis of a cellulose grafted polyacrylamide/hydroxyapatite composite hydrogel through a suspension polymerization method. As shown in Fig. 12, the HAp material were embedded in the hydrogel matrix through ionic crosslinking of species OH^-, Ca^{2+}, $PO_4{}^{3-}$ and amide groups of acrylamide and/or hydroxyl groups in the cellulose backbone. The prepared hydrogel exhibited a maximum adsorption capacity of 175 mg/g for removal of copper ions from an aqueous solution.

The Table 1 summarizes some hydroxyapatite-based materials for heavy metals uptake from aqueous solution with maximum adsorption capacity. Most studies on HAp-based materials are at a preliminary stage of research, and there is a need of research effort of this class of materials for actual wastewater containing multiple pollutants.

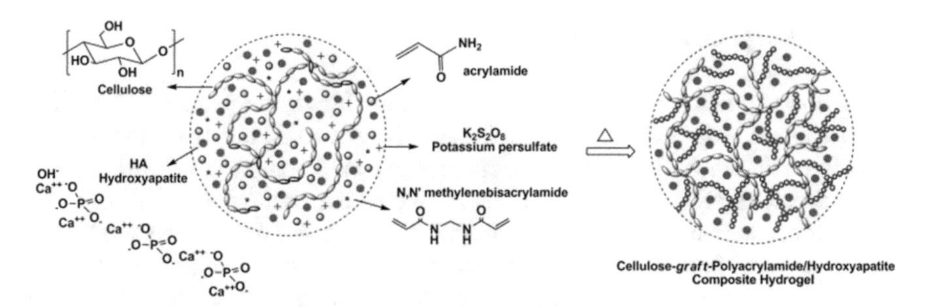

Fig. 12 Synthesis of cellulose grafted polyacrylamide/hydroxyapatite composite hydrogels. Reproduced with permission from [80]. Copyright 2013 Elsevier

Table 1 Hydroxyapatite based materials for heavy metals removal

Adsorbent HAP-based materials	Adsorbate	Maximum Adsorption Capacity ($mg{\cdot}g^{-1}$)	Refs.
HAP	Ni^{2+}	46	[82]
HAP	Ni^{2+}, Co^{2+}	186.6, 22,5	[83]
Nano-HAP (n-HAP)	Pb^{2+}	186.48	[68]
Mesoporous HAp	Pb^{2+}, Cd^{2+} and Ni^{2+}	181, 122, 81	[84]
Short rod HAP (rHAP)	$Pb2^{+}$	460.8	[64]
Flake-flower HAP (fHAP)	Pb^{2+}	188.7	[64]
Dandelion HAP (dHAP)	Pb^{2+}	819.7	[64]
Stoichiometric HAP (S-HAP)	Pb^{2+}, Cu^{2+}	81, 19	[67]
Porous HAP	U^{6+}	111.4	[85]
Nano Esterified HAP (n-EHAP)	Pb^{2+}	2397	[71]
Sulfonated-HAp (BDS-HAp)	Zn^{2+}, Cd^{2+}, Pb^{2+}	275.25, 354.09, 507.64	[74]
Soledronate functionalized HAP (ZA-HAP)	Pb^{2+}, Cu^{2+}	1460.14, 226.33	[70]
Alendronate doped HAP (AL-HAP)	Pb^{2+}, Cd^{2+}, Cu^{2+}	1431.8, 469.0, 226.6	[75]
Organophosphonate-modified HAp (NTP-HAP)	Pb^{2+}, Zn^{2+}	450, 300	[73]
Carboxylmethylcellulose@HAP (CMC@HAP)	Pb^{2+}	625	[81]
Chitosan@HAP (CTS@HAP)	Pb^{2+}	909	[81]
Fe_3O_4/hydroxyapatite/β-cyclodextrin	Cd^{2+}, Cu^{2+}	100.00, 66.66	[86]
Cellulose-graft-polyacrylamide/HAP hydrogel	Cu^{2+}	175	[80]
Magnetic HAP/Agar composite beads	Pb^{2+}, Co^{2+}, Cu^{2+}	842.6, 105.1, 71.6	[87]

4.1.2 Adsorption of Fluoride

In addition to pollution of water by phosphate and nitrate, the presence of fluoride in water poses serious risks to human because long-term up taking of fluoride can result in numerous health problems including as bone fluorosis. The World Health Organization (WHO) classified fluoride as one of the contaminants of water for human consumption [88]. To this end, the WHO established a strict standard for the concentration of fluoride in drinking water of not exceeding 1.5 mg/L [89, 90].

Among of various reported materials used for fluoride adsorption, hydroxyapatite based materials has attracted much interest due to their stability, non-toxicity and biocompatibility without unfavorable effect on water quality. In addition, HAp can be easily synthesized from natural phosphate and animal waste materials, which makes it economically practical and vastly available. In this context, Smant et al. [91] successfully synthesized HAp using *Limacine artica* shell as the starting material. The as-prepared HAp material exhibited an adsorption capacity of 28.57 mg/g toward F^- ion adsorption. The initial pH of solution played an important role on removal efficiency of fluoride (Fig. 13). Consequently, in a high acidic medium the removal efficiency of fluoride was found to be insignificant because the high concentration of H^+ favors the formation weakly ionized HF, subsequently generating HF_2^- and H_2F^+ through homoassociation ($3\ HF \rightarrow HF_2^- + H_2F^+$). In addition, below pH 4.8 the formation of soluble $Ca(OH)_2$ and a stable di-calcium phosphate in aqueous phase decreased the amount of solid HAp. The increase of pH from 2.1 to 6.3 has positive impact on fluoride removal efficiency. HAp adsorption been purely electrostatic in nature, this result can be explained by interaction of fluoride ion with protonated OH groups of HAp ($F^- \cdots OH_2^+$ interaction). However, the increase in pH below six

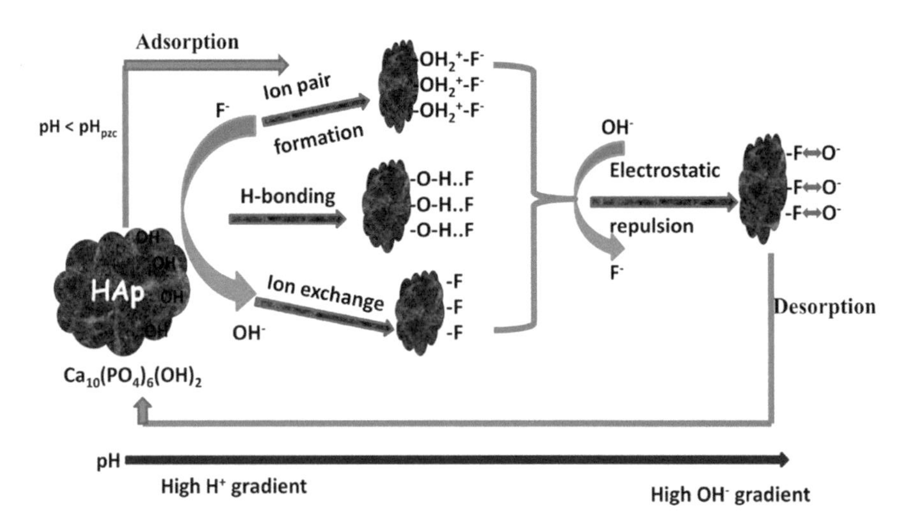

Fig. 13 Adsorption mechanism of F^- ions using HAp. Reproduced with permission from [91]. Copyright 2017 Elsevier

resulted in a decrease of fluoride removal efficiency. This result confirm that the adsorption process did not depend merely on the surface charge of HAp, which was positive at pH below pHzpc 8.128 due to the prevalence of $=CaOH_2^+$ species and negative above pHzpc due to the predominance of $=PO^-$ species. The driving force for adsorption of a negatively charged fluoride ion on HAp surface could be attributed to several factors like, ion pair formation ($F^-\ldots OH^{2+}$), H-bonding ($F^-\ldots H\ldots O$) and ion exchange (F^- will exchange OH^-).

In another study, Tomar et al. [92] reported the synthesis of a hybrid adsorbent based on hydroxyapatite modified activated alumina (HMAA) prepared the dispersion of HAp nanoparticles inside activated alumina granules. The experiments of this study was carried in batches and as well as column mode. The maximum adsorption capacity of HMAA composite was 14.4 mg/g, much higher than that of activated alumina, demonstrating a synergy effect between HAp and activated alumina toward fluoride removal from contaminated drinking water. More importantly, the adsorbent exhibited a potential for reuse over multiple cycles without any appreciable loss in its removal capacity over repeated use.

In fact, the application of HAp in adsorption columns is limited because of its powder form. In recent years, an efficient approach to synthesize new composites to overcome the above issues has been achieved, by integrating inorganic HAp material with other functional polymeric materials, which have synergistic advantages toward fluoride adsorption. For example, Pandi et al. [93] synthetized HAp encapsulated with alginate polymer to form beads hydrogel. Firstly, the authors prepared HAp powder through co-precipitation method. Subsequently, the prepared HAp material and alginate polymeric gel were fully mixed together and then kept for 3 h under stirring. The homogeneous HAP-Alg mixture was dropped into $LaCl_3$, $7H_2O$ solution for crosslinking. The as-prepared composite exhibited adsorption capacity of 3.72 mg/g for fluoride uptake. In addition, the authors study the effect of presence of other anions in solution like Cl^-, NO_3^{2-}, SO_4^{2-}, and HCO_3^-. The mentioned results demonstrated that the presence of anions like like Cl^-, NO_3^{2-} and SO_4^{2-} would not have a significant negative effect on the fluoride uptake capacity. However, the presence of HCO_3^- ions decreased the adsorption capacity, because the increase in solution pH simultaneously reduces the active sites for fluoride sorption.

In summary, Table 2 presents the maximum adsorption capacity, and fluoride concentration range for the adsorption of fluoride by hydroxyapatite based materials.

4.1.3 Adsorption of Organic Pollutants

Toxic organic pollutants such as dyes, phenols, antibiotics, pesticides and emerging pollutants are discarded into wastewater as byproducts of several industrial activities. These pollutants are toxic, carcinogenic and cause several environmental problems to our environment. Figure 14 represents the chemical structure of some organic pollutants. Different treatment methods were applied for the removal of different organic pollutants in water such as coagulation, precipitation-flocculation, membrane filtration, reverse osmosis, ozonation, etc. These technologies assure great results,

Table 2 Hydroxyapatite based materials for fluoride removal

Adsorbent HAP-based materials	Fluoride concentration range (mg/L)	pH	Maximum adsorption capacity (mg g^{-1})	Refs.
HAp derived from Limacine artica shells	10	6.3	28.57	[91]
HAp derived from egg shells	10	6	22.3	[94]
HAp derived from phosphogypsum	10–50	2–11	19.74 – 40.81	[95]
nanosized HAp	50	5	3.44	[95]
non-calcined synthetic HAp	5	7	4.38	[96]
Alumina-modified hydroxyapatite	5–10	7	32.57	[92]
CTAB coated HAp powder	10	3	9.36	[97]
cellulose@HAp nanocomposites	10	6.5	4.2	[98]
Alginate@HAp composites	10	5	3.72	[93]
Chitosan@HAp composites	10	3	1.56	[99]
HAp/Chitin composites	10	7	2.84	[100]

but limited efficiency in the removal of organic pollutants at low concentrations [101]. In this context, adsorption process have been extensively investigated for water remediation with low concentration of pollutants. In addition, the adsorption is a low cost technique, simple and energy considerations [102]. In this section, we will discuss and compare recent developments on the adsorption organic pollutants from water using HAp and HAp-based materials.

Organic Dyes

Water pollution by organic dyes has set off a universal threat because they are damaging the aesthetic nature of the environment. In addition, these toxic substances can damage fish and wildlife, resulting in long-term ecological effects. Various materials have been studied to remove dyes from wastewater such as porous carbon, clay, zeolite and metals oxides [103–105]. Hydoxyapatite based materials have also been investigated in the adsorption of different cationic and anionic dyes. Regarding Congo red (CR) adsorption, Chahkandi et al. [106] reported the elaboration of hydroxyapatite nanoparticles via alkoxide-based sol–gel method, which showed an efficient adsorption of CR from aqueous solution (Fig. 15a). The as-prepared material exhibited maximum adsorption capacity of 487.80 mg/g, which was probably due to

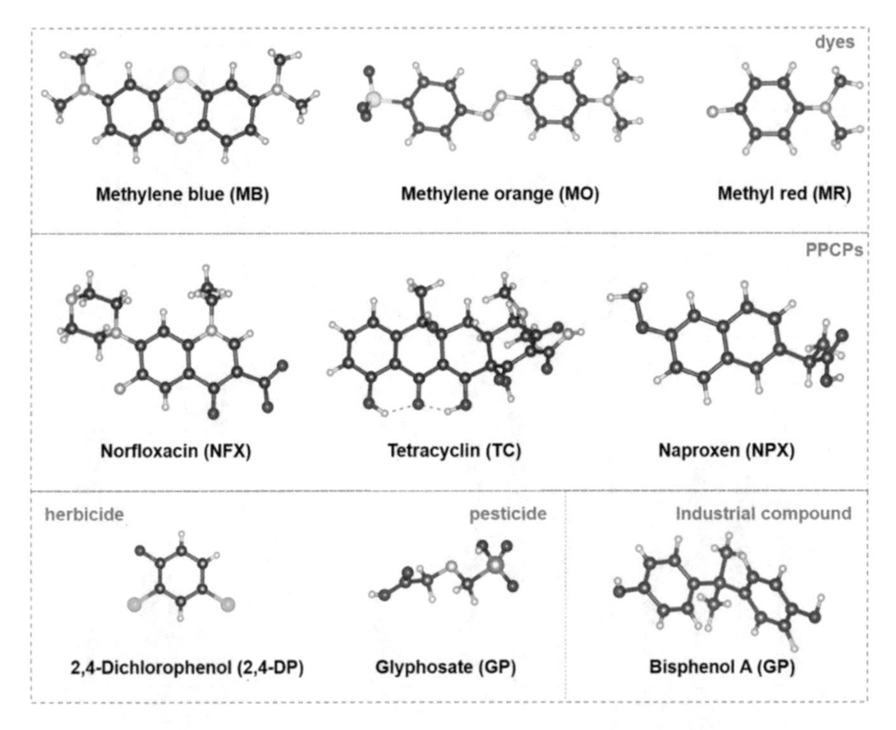

Fig. 14 Chemical structures of the most pollutants removed from water by HAp based materials. Reproduced with permission from [102]. Copyright 2019 American chemical society

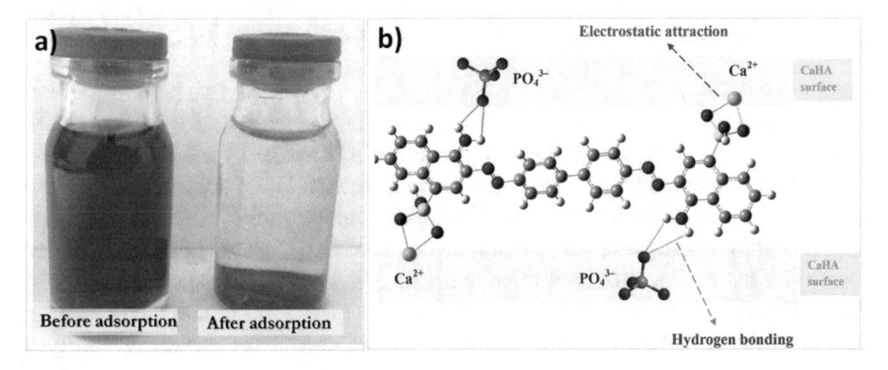

Fig. 15 **a** demonstration of dye adsorption, **b** possible theoretical mechanism for CR adsorption by HAp. Reproduced with permission from [106]. Copyright 2017 Elsevier

the electrostatic interaction between the positive charge of Ca^{2+} ion and negatively charged SO_3^{2-} groups of Congo red molecule as well as hydrogen bonding of amine group and phosphate groups in surface of HAp (Fig. 15b).

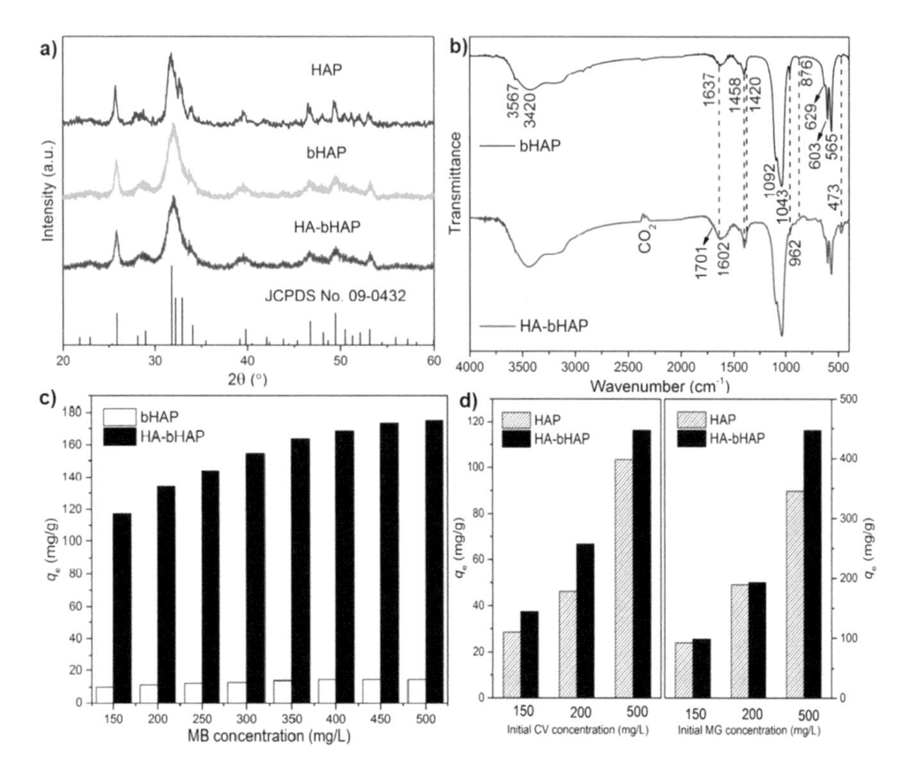

Fig. 16 **a** XRD patterns and **b** FTIR spectroscopy of prepared biogenic hydroxyapatite (bHAP), humic acid-impregnated bHAP (HA-bHAP) and synthetic HAP (HAP), Adsorbed amount of **c** methylene blue **d** crystal violet and malachite green using HA-bHAP compared with bare bHAP. Reproduced with permission from [108]. Copyright 2016 Royal society of chemistry

In another study, Mehri et al. [107] reported the preparation of hybrid materials by grafting biogenic monoamine in the surface of HAp. The hybrid Tyramine-HAp material were prepared by in situ hydrothermal method in the presence of different amounts of tryamine molecules. The FTIR spectroscopy analysis of HAp and modified HAp showed the presence of vibration bands of PO_4^{3-} group of apatite and new absorption bands was observed after tyramine functionalization. In addition, from the XRD patterns, the hexagonal crystal structure of HAp was found to be preserved after tyramine grafting. However, a loss of crystallinity with increasing the tyramine amount was observed. Furthermore, solid-state NMR spectroscopy analysis was given in the paper to further confirm the successful grafting of tyramine in the apatite structure. The reported ^{31}P NMR-MAS results demonstrated the presence of tyramine did not alter the crystallographic site of apatite phosphorus (PO_4^{3-}). In addition, using ^{1}H MAS-NMR they observed the presence of additional resonance signals at 1.35 ppm (not detected in the HAp), which represents signal from proton of $-CH_2$ groups of tyramine, added in matrix after HAp surface grafting. The reported hybrid organoapatite material (2Tyr-HAP) with 4.8% weight of Tyramine showed maximum adsorption capacity of 12.55 mg/g for removal of MO in aqueous solution.

In another work, Wang et al. [108] described a combination of humic acid with biowaste-derived HAp for highly efficient removal of methylene blue (MB) from aqueous solution. The HAp was prepared from eggshell waste as calcium source. The prepared HAp material was then impregnated with humic acid (HA) at pH 5 through adsorption process. The amount humic acid loaded onto HAp was calculated and found to be 118.7 mg/g. As presented in Fig. 16a , the XRD patterns of synthetic HAP, biogenic HAp (bHAP) and humic acid modified bHAP (HA-bHAP) show all characteristic peaks of hexagonal HAp crystal structure. The biogenic HAp exhibited a poor crystallinity compared to synthetic HAp, which probably due to the carbonate incorporation in bHAP. The FTIR spectra of bHAP and HA-bHAP (Fig. 16b) displays all characteristic peaks for PO_4^{3-} (473, 565, 603, 962, 1043 and 1092 cm^{-1}), OH^- (629 and 3567 cm^{-1}), adsorbed water (1637 and 3420 cm^{-1}) and CO_3^{2-} ions (876, 142 and 1458 cm^{-1}) which indicated the incorporation of carbonate ion in synthesized pure HAP. Additionally, in the case of humic acid modified bHAP, new additional peaks at 1608 and 1701 cm^{-1} were observed, which assigned to C=O vibration bond in carboxylic salt and free carboxylic acid, respectively. Authors investigated the comparison of bHAP and HA-bHAP for MB removal (Fig. 16c). The results showed that HA-bHAP exhibited higher adsorption capacity (131.72 mg/g) towards removal of 200 ppm of MB compared to unmodified bHAp, which was only 14.27 mg/g. The difference of the adsorption capacity HAP and HA-bHAP was probably due to negative charge of HAp surface after modification with humic acid. The maximum adsorption capacity of MB using HA-bHAP was 393.47 mg/g. Furthermore, the feasibility for adsorption of other cationic dyes such as malachite green and crystal violet using HA-bHAP was also investigated. The results in (Fig. 16d) indicated that HA-bHAP also exhibited excellent performance for these cationic dyes.

Guan et al. [109] developed polyalcohol-coated HAp nanocomposite from D-fructose-1,6-phosphate trisodium salt octahydrate (DFP) and calcium nitrate via simple hydrothermal process. The XRD pattern showed hexagonal HAp as unique crystalline phase (Fig. 17**a**). The reported ^{13}C NMR CPMAS results (Fig. 17b) demonstrated the opening of the fructose rings and formed mostly alcohol groups as proven by the absence of a 103 ppm signal and the presence of peaks in the range 60–80 ppm. In addition, the reported ^{31}P NMR CPMAS (Fig. 17c) show one isotopic signal at 2.2 ppm suggesting the presence of one that the phosphate groups of DFP in the prepared HAp. Based on these analyses, the authors schematized the polyalcohol-coated HA nanocomposties as represented in Fig. 17d.

The produced Hap-based adsorbent was evaluated to remove methyl orange, congo red and methyl blue, from water. Based on the Langmuir model, the maximum adsorption capacity towards methyl orange, congo red and methyl blue was 14.7, 170.7 and 379.1 mg/g, respectively. The high adsorption capacity could also be attributed to the high surface area of composite (203.2 m^2/g).

In summary, Table 3 provides the adsorption capacity of various HAP adsorbents for the removal of different dyes.

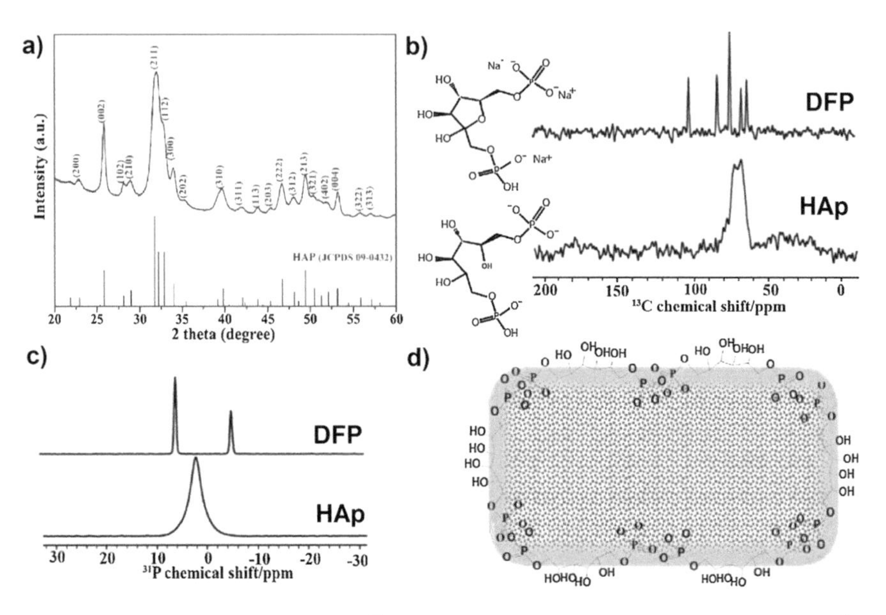

Fig. 17 **a** XRD pattern of HAp and **b** ^{13}C cross polarization magic angle spin (CPMAS) **c** ^{31}P CPMAS NMR spectra of DFP and HAp **d** the schematic diagram of polyalcohol-coated HAp. Reproduced with permission from [109]. Copyright 2018 Elsevier

Table 3 Hydroxyapatite based materials for dyes removal

Adsorbent HAP-based materials	Dye	Type of Dye	Maximum Adsorption capacity (mg g $^{-1}$)	Refs.
HAP	Congo red	Anionic	487.80	[106]
Magnetic carbonate HAp/graphene oxide	Methylene blue	Cationic	546.40	[110]
Tyramine-HAp	Methyl orange	Anionic	12.55	[107]
Hide substance/chitosan/HAp	Methylene blue, sunset yellow, Orange-NR, RedVLN, Blue-113, Green-PbS	Cationic, industrial dyes	3.8, 168, 496, 477, 488, 274	[108]
Humic acid modified biogenic HAp	Methylene blue	Cationic	393.47	[111]
Nano-hydroxyapatite polymeric hydrogels	Acid Blue 113	Anionic	29.52	[112]
HAp derived from egg shells	Reactive Yellow 4	Anionic	127.9	[113]
polyalcohol-coated hydroxyapatite	Methyl orange, Congo red, Methyl blue	Anionic, Anionic, Anionic	14.7, 170.7, 379.1	[109]

Phenolic Compounds

Phenol and phenolic compounds generated from the discharge of chemical industries are very toxic even at low concentration. Hence, its removal from waste water is considered as a prime filed of research, especially with the existing environmental laws [7]. In this context, numerous technologies have been considered to eliminate these toxic pollutants from water (i.e. oxidation with ozone/hydrogen peroxide, coagulation-flocculation, biological methods, electrochemical oxidation, reverse osmosis and photocatalytic degradation) [114]. Among these technologies, adsorption process remains the best, as it can generally remove all types of phenols, and the effluent treatment is convenient because of its simple design and easy operation [115]. Strong research efforts have been devoted to materials adsorbent for phenol contaminated water treatment. In particular, hydroxyapatite has been proposed for the removal of a wide range of phenolic compounds in water. Lin et al. [116] investigated the removal of phenol in aqueous solution using HAp nanopowders. The study was carried out at several pH values. The increase of pH to high-acidity or to high alkalinity resulted in the increase in the phenol adsorption capacity. In a low pH values, the HAp would be protonated and became positive which led to donor–acceptor interactions between the aromatic rings of the phenol. However, at high pH values of alkalinity, the phenol would be ionized in solution and this led to increase in the ionic strength. The maximum phenol adsorption capacity was obtained as 10.33 mg/g for 400 mg/l initial phenol concentration at pH of 6.4. In another report, Bouiahya et al. [117] prepared alumina-hydroxyapatite (Al-HAP) composites via dissolution–precipitation of natural phosphate. Different amounts of aluminum precursor was added to the mixture of Ca^{2+} and H_3PO_4 prepared from dissolution of natural phosphate, and the pH of solution was adjusted to 10 by ammonia solution NH_4OH. After preparation and characterization of Al-HAP materials, the authors evaluated its adsorption capacity to remove phenol in aqueous solution. The monitoring of adsorption was carried out using UV–visible spectroscopy at $\lambda_{max} =$ 273 nm. As shown in Fig. 18a, the HAP exhibited a lower adsorption capacity for removal of phenol; however, the phenol adsorption capacity increased with increase of Al^{3+} content in the composites. These results can be explained by the change of surface charge after adding of Al^{3+}. As schematized in Fig. 18b, the adsorption of phenol on Al-HAp composites was due to the electrostatic interaction/or hydrogen bonding between the negatively charged oxygen atoms of the OH groups in phenol or π-electrons of the aromatic nucleus and the protonated alumino groups ($-AlOH_2^+$) of Al–HAP composites. The maximum adsorption capacity 56.43 mg/g of phenol was obtained using 20Al-HAp composite with an amount of 11.4% of Al^{3+}. Table 4 recapitulate a few studies dealing with the adsorption of phenolic compounds using HAP based materials.

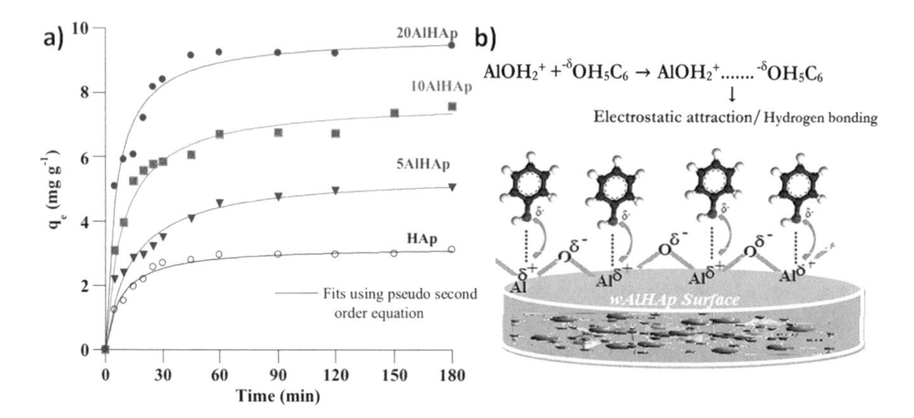

Fig. 18 **a** Effect of contact time on the adsorption of phenol on the wAlHAp composites, **b** Scheme of electrostatic interactions between phenol and wAlHAp surface. Reproduced with permission from [117]. Copyright 2019 Elsevier

Table 4 Hydroxyapatite based materials for phenol removal

Adsorbent HAP-based materials	Phenolic compound	Maximum adsorption Capacity (mg g $^{-1}$)	Refs.
HAP	Phenol	10.3	[116]
Ba substituted HAP	Phenol	220	[118]
Al_2O_3-HAP	Phenol	56.43	[117]
Cellulose nanofibrils/HAP films	Phenol	64	[115]
Fe_3O_4@HAP	Phenol	18	[119]
Al_2O_3-HAP	2-cholorophenol, 2-nitrophenol	9.59, 5.33	[120]
Ceramic HAP foam	Bisphenol A	76.5	[121]

Emerging Organic Pollutants

Water pollution by various emerging organic pollutants has proven to be a serious environmental concern, since they may cause ecological or human health impacts [122]. These emerging pollutants are persistent with high stable structure. The common emerging organic pollutants are cosmetics, herbicides, pesticides and pharmaceutical compounds such as antibiotics and drugs [122]. Among the different emerging organic pollutants, the elimination of some commonly found compounds in water have been studied using HAp based materials. For example, Li et al. [123] reported Fe incorporated hydroxyapatite (Fe-HAp) for the removal of tetracycline in water. Structural characterizations indicated the efficient incorporation of iron in HAp structure. In comparison with pure HAP, the Fe-HAp with different amounts of iron exhibited higher adsorption capacity for the removal of tetracycline from

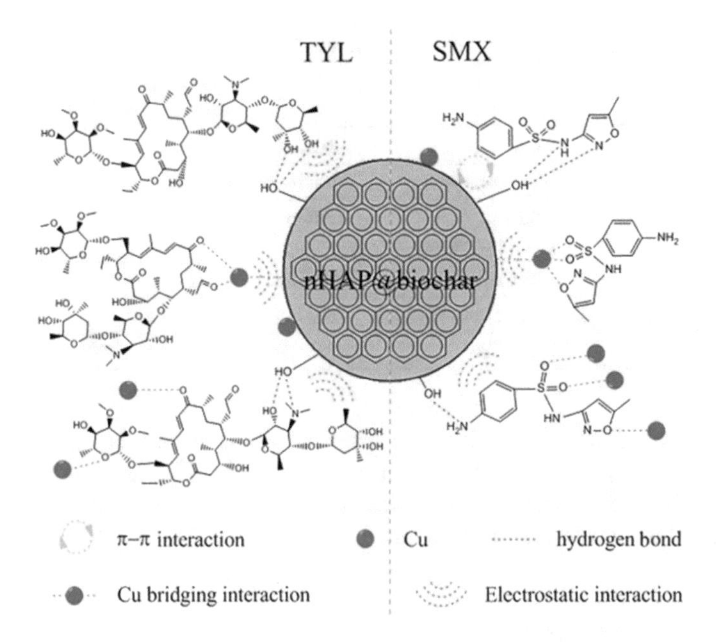

Fig. 19 Schematic illustration of TYL/SMX and Cu adsorption mechanism by nHAP@biochar. Reproduced with permission from [124]. Copyright 2020 Elsevier

aqueous solutions. The maximum adsorption capacity of 45.39 mg/g was obtained when 0.02Fe-HAP was used. The authors stated that the interaction of Fe-HAp with tetracycline in the presence of humic acid was important for the efficiency of adsorption. The presence of humic acid in solution improved the adsorption through hydrogen bonding between carboxylic, phenolic hydroxyl groups of humic acid and the electron-donors (–OH/O^-, –NH_2) of tetracycline. In addition, the presence of positive surface charges of Fe^+ species in HAp involves surface complexation with the negative charges of the tetracycline. In another study, Li et al. [124] reported the preparation of biochar stabilized by nano-hydroxyapatite (nHAP@biochar) toward coadsorption of tylosin/sulfamethoxazole (TYL/SMW) and Cu $^{2+}$ in aqueous solution. The adsorption capacity for tylosin and sulfamethoxazole was around 160 mg/g and 140 mg/g, respectively. The authors used FTIR spectroscopy and XPS analysis to study the mechanism of adsorption. Figure 19 illustrate the adsorption mechanism, which was as follows:

- H-bonding may be the main interaction of TYL and SMX with nHAP@biochar and weak π-π interactions for SMX.
- During adsorption, TYL/SMX-Cu complexes were formed.
- TYL might act as a bridge between Cu(II) and nHAP@biochar.
- Cu(II) might act as a bridge between SMX and nHAP@biochar to form nHAP@biochar-TYL-Cu and nHAP@biochar-Cu-SMX.

Table 5 Hydroxyapatite based materials for removal of emerging organic pollutant

Adsorbent HAP-based materials	Emerging organic pollutant	Type	Maximum adsorption capacity (mg g^{-1})	Refs.
Fe-HAp	Tetracycline	Antibiotic	45.39	[123]
Nano-HAP	Oxytetracycline	Antibiotic	0.0326	[125]
nHAP@biochar	Tylosin /sulfamethoxazole	Antibiotic	160, 140	[124]
HAP/Clay	Tetracycline	Antibiotic	76.02	[126]
Biomorphic nano-HAP	Triclosan, Ofloxacin	Antibiotic	133.3, 29.15	[127]
Nano-HAP	Atrazine	Pesticide	0.00995×10^{-10}	[128]

Similar adsorption mechanism was also obtained by et Yoan et al. [125] for co-adsorption of oxytetracycline (OTC)/ metallic ion species onto nano-hydroxyapatite (nHAP). Table 5 provides some studies dealing with the adsorption of emerging organic pollutant using HAp adsorbents.

4.1.4 Oil/Water Separation

Oil/water mixtures' management is considered as a challenging task, from an academic and social stand point. The concerns regarding the environment and energy have captured a global consciousness owing to the polluted environment especially the aquatic one, ruined aquatic ecosystems and energy requests from petroleum exploitation and reprocessing [129–131]. Numerous oil spill accidents have taken place in the last decade causing catastrophic repercussions on the aquatic ecological system and environment, such as the Gulf of Mexico in 2010 and Qingdao, China in 2013, etc. Furthermore, other industries also take part of this pollution owing to the production of oily wastewater, especially in the form of oil/water emulsions, such as pharmaceutical, chemical, petrochemical, metallurgical, food, textile and leather industries [132, 133].

Generally, The standard methods to overcome oil spills are mainly focused on the utilization of emulsion breakers, chemical dispersants, in situ burning, air flotation, ultrasonic separation, coagulation and so on [134, 135]. Among them, adsorption/filtration techniques are considered as one of the most efficient approaches to separate oil/water mixtures. Subsequently, the R&D aiming to explore highly efficient oil/water separation materials have become an urgent need. In this context, oil adsorption materials with multifunctional properties have drawn extensive interest in recent years. By controlling the surface wettability toward water and oil, high separation efficiencies of superhydrophobic or superoleophilic materials can be achieved [136–138]. From this angle, hydroxyapatite and nanostructured hydroxyapatite-based materials as inorganic materials have attracted great attention in this field of water decontamination offering interested properties, compared with organic materials,

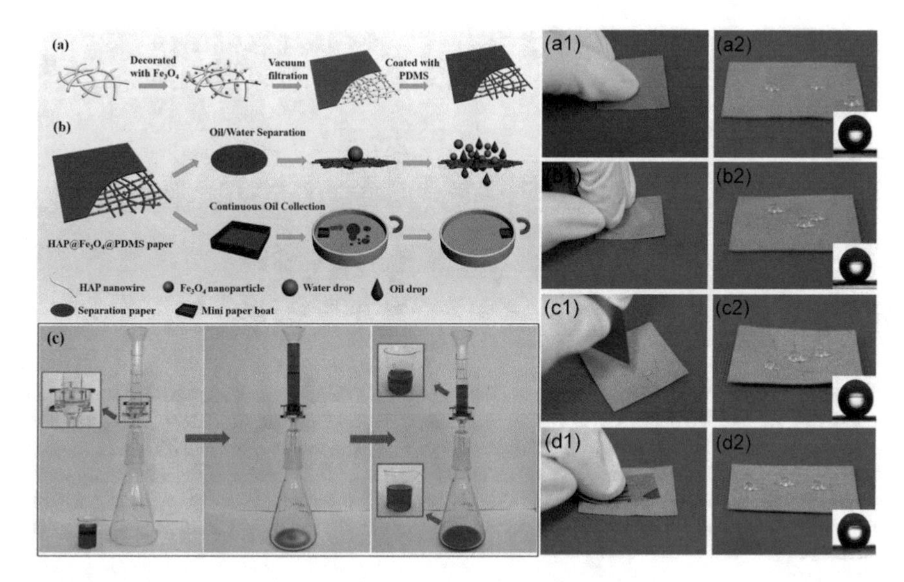

Fig. 20 **a** Illustrative representation of HAP/Fe_3O_4/PDMS paper processing, **b** Oil/water separation over the elaborated HAP/Fe_3O_4/PDMS, **c** Oil/water separation experiments over HAP/Fe_3O_4/PDMS (red: chloroform and blue: water). HAP/Fe_3O_4/PDMS paper superhydrophobicity stability following physical scarring: **a1** wipe using a finger, **b1** peeling using a tape, **c1** scratch with a blade, **d1** abrasion. **a2, b2, c2, d2** Stability of water contact angle following physical scarring. Reproduced with permission from [143]. Copyright 2018 American Chemical Society

including chemical stability, high thermal stability, fire resistance and infrequently releasing toxic gas, etc. [139–142].

For instance, Yang et al. [143] reported the preparation of hydroxyapatite (HAP)/Fe_3O_4 nanoparticles modified with polydimethylsiloxane (PDMS) deposited on paper which endow this later magnetic, fire retardancy and superhydrophobic features (Fig. 20a, b). The prepared HAP/Fe_3O_4/PDMS paper exhibited promising fire resistance, thermal and superhydrophobic stability (Fig. 20a1–d2). Furthermore, the porous morphology and superhydrophobicity of the produced HAP-base paper showed selective oil infiltration, high permeation flux and good recyclability (Fig. 20c). In addition, the produced HAP/Fe_3O_4/PDMS can act as a collecting tool that can be magnetically driven to oil-polluted areas where it can be recovered easily by a magnet.

In another study, Li and coworkers [144] proposed a route for the elaboration of Kevlar fiber (KF) incorporated paper (SH) based on nanostructured hydroxyapatite nanowires (HAPNWs), followed by its modification using long chain acid which is stearic acid (STA). The KF and HAPNWs form a 3D porous morphology, thus immobilizing TiO_2 nanoparticles, consequently enhancing the flexibility and mechanical strength of the prepared composite paper. The produced HAP-based paper was used to efficiently separate oil/water mixed solution owing to its supeoleophilicity. The

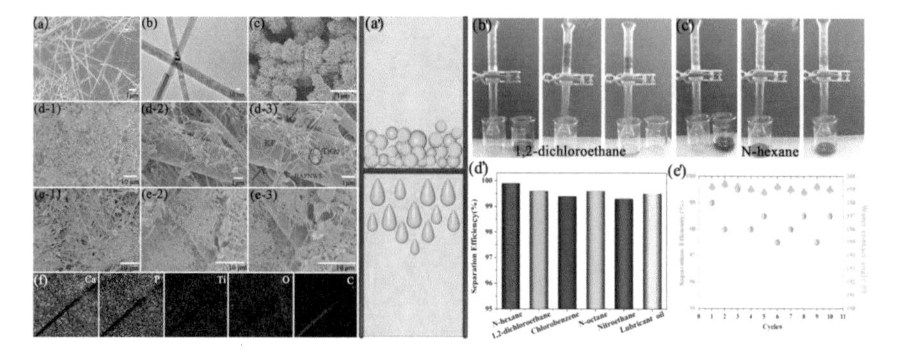

Fig. 21 **a** neat HAPNWs SEM micrograph, **b** neat HAPNWs TEM image, **c** nano-TiO_2 SEM micrograph, **d, e** unmodified paper and SH paper SEM micrographs. **f** unmodified paper EDX mapping. **a′** Oil/water separation scheme, **b′** dichloroethane/water separation experiments (dichloroetheane: blue), **c′** N-hexane/water separation experiments (N-hexane: red), Different oil/water mixture separation over SH paper, **e′** recyclability of the SH paper after several separation experiment with dichloroethane/water. Reproduced with permission from [144]. Copyright 2019 Royal society of chemistry

high and efficient oil/water separation was owned to the plentiful nanopores, excellent capillary action and superhydrophobic properties of produced composite paper (Fig. 21).

With the same state of mind, Lu and collaborators [140] reported an approach for the manufacturing of highly hydrophobic (contact angle of 137°) and nonflammable inorganic HAP-based paper via the preparation of nanostructured long HAP nanowires. The high aspect ratio nanowires were elaborated using a precursor which is calcium oleate. The produced HAP paper demonstrated interesting features such as flexibility and non-flammability. The elaborated HAP nanowires-based paper displayed promising performances in oil/water separation purposes; it exhibits high adsorption capacities for various organic pollutants, for example, the HAP nanowiers-based paper of revealed a high adsorption capacity of 7.3 g g^{-1} for chloroform (Fig. 22).

In the same manner, Chen et al. [141] reported the successful manufacturing of sodium oleate modified hudrophobic HAP nanowires. They stated that the HAP nanowires assembled into self roughened fibers during the filtration process, thus constructing a layered structure on the paper. They also demonstrated that obtained layered structure preserved the superhydrophobic feature owing to the enhancement of the resistance to physical damage. The prepared HAP-based paper displayed repellency to numerous liquids and self cleaning characteristic. Moreover, the authors explored the potential application of the produced HAP-based paper as a effective membrane for oil/water separation (Fig. 23).

Furthermore, Inorganic-based aerogels have aroused great interest owing to their numerous qualities such as low cost, biocompatibility, non toxicity, etc. Generally, aerogel-based materials offer outstanding features such as high porosity, low density, and chemical inertness which make them promising candidate for oil removal from

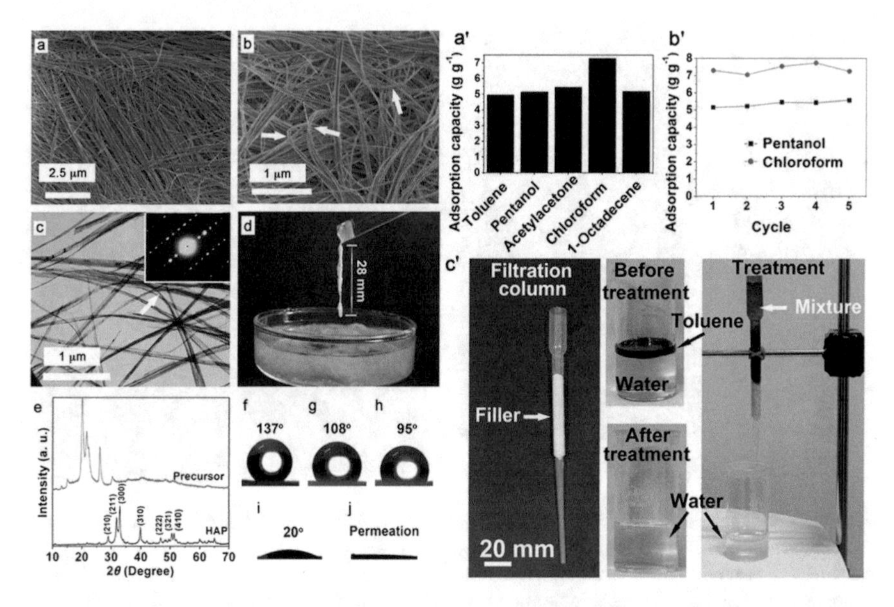

Fig. 22 **a, b** HAP nanowires SEM images. **c** HAP nanowires TEM image. **d** HAP macro-fiber obtained from a suspension of HAP wires in ethanol. **e** Calcium oleate XRD diffractogram. **f–j** HAP samples water contact angle photos processed in different reaction conditions. **a′** HAP paper's adsorption capacities towards numerous organic pollutants. **b′** HAP paper recyclability following pentanol and chloroform adsorption. **c′** Toluene/water separation experiment using HAP paper. Reproduced with permission from [140]. Copyright 2014 Wiley and Sons

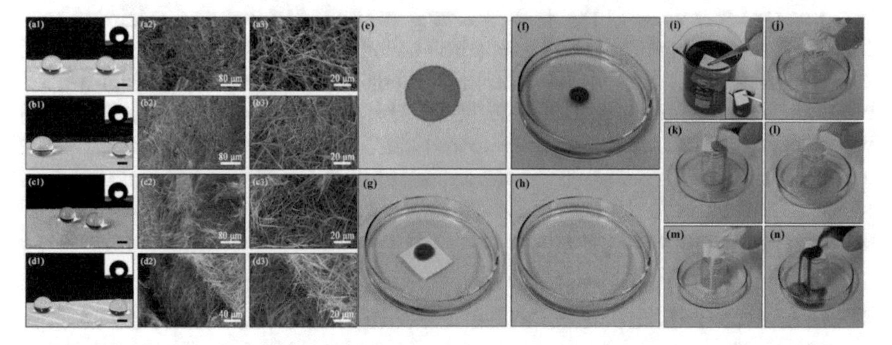

Fig. 23 **a1–d1** water contact angle of sodium oleate modified HAP-based paper after different physical destruction showing the stability of the deposited layered structure. **a2–d3** SEM images of the produced papr after different mechanical damages. **e–i** Oil/water separation experiment (cyclohexane: red, water: blue,). **j–n** Repellency to different liquids including mineral water, tea, juice, milk and coffee, respectively. Reproduced with permission from [141]. Copyright 2016 American Chemical Society

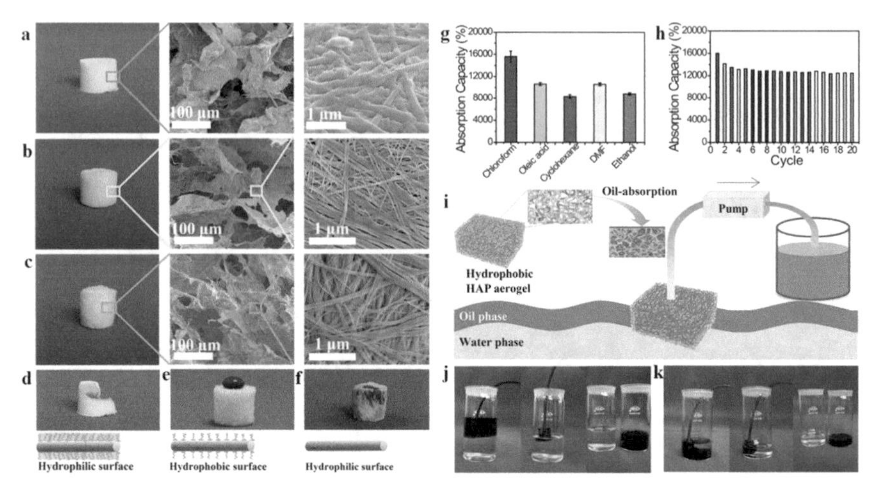

Fig. 24 **a–c** Freeze-dried HAP samples' digital and SEM images. **d–f** HAP sample illustrative representation. **g** solvent/oil uptake of the prepared HAP aerogel for different oils and organic solvents. **h** Reusability of HAP aerogel for chloroform uptake. **i** illustrative representation showing a montage of dynamic oil/water separation using HAP-based aerogel. **j–k** dynamic separation of cyclohexane (left) and chloroform (right). Reproduced with permission [145]. Copyright 2018 American Chemical Society

oily water. Nevertheless, in most cases, they often possess poor mechanical properties mainly a brittle characteristic, thus restricting their real application as membranes of oil/water separation. In this matter, Zhang et al. [145] developed calcium oleate modified hydroxyapatite (HAP) nanowire –based aerogel with excellent elasticity, high porosity (porosity $\approx 99.7\%$), ultralight (density 8.54 mg/cm^3), and highly adiabatic (thermal conductivity 0.0387 W/m K). The elaborated modified HAP aerogel revealed great potential in oil/water separation. The elaborated hydrophobic HAP aerogel displayed high absorption capacities for different oils and organic solvents ranging from 83 to 156 times the weight of the aerogel with recyclability up to 20 cycles. The authors supported these results by two major explanations:

- The hydrophobic nature induced by the calcium oleate modification
- The presence of high porosity which gave birth to interconnected porous network that furnishes a substantial capacity for oil/organic solvent absorption (Fig. 24).

4.2 *Catalysis for Organic Pollutants Degradation in Water*

Recently, oxidation processes including sulfate radical-induced degradation, photocatalysis, and Fenton reactions, using HAp based materials have attracted a lot of scientific interest due to their high efficiency and reproducibility. The photocatalysis processes initiated by single electron transfer resulting in the creation of reactive species ($^{\bullet}O_2$, $OH^{\bullet}$, and h^+). In addition, hydrogen peroxide (H_2O_2), persulfate (PS) and peroxymonosulfate are also used in combination with HAp based materials to produce highly reactive species ($SO_4^{\bullet-}$ and $OH^{\bullet}$) to boost organic pollutants' degradation and mineralization. In this segementnt, we're going to be interested in the comparison between different HAp based materials concerning organic pollutants degradation in water.

4.2.1 Catalyst for H_2O_2 and Peroxymonsulfate Activation

A very few studies reported the degradation of organic pollutants by catalytic activation of H_2O_2 and PMS. For example, Das and coworkers [146] reported the preparation of Ni/HAP/$CoFe_2O_4$ composite for degradation of organic dyes (methylene blue and methyl orange) in the presence of H_2O_2. The magnetic composite showed a degradation efficiency of 90% (methyl orange) and 99% (methylene blue) in 90 min. in this study, the nickel supported on the hydroxyapatite play important role for generation of radical species ($HO^{\bullet}$ and $HOO^{\bullet}$). The same group also reported the preparation of Zn^{2+} supported onto the magnetic hydroxyapatite (Zn/HAP/$MgFe_2O_4$). The composite exhibited high catalytic activity for degradation of malachite green in the presence of H_2O_2 [147]. In the presence of H_2O_2, a removal efficiency (100%) of malachite green was obtained in 2 min. Pang et al. [148] synthesized cobalt doped hydroxyapatite via ion exchange method followed by calcination at 500 °C. The as-prepared material was evaluated for PMS activation for Rhodamine B degradation. The XRD result confirmed that Co species existed in HAP lattice structure and in form of Co_3O_4 at HAP surface. The Co-HAP catalyst with 2% of cobalt species showed a degradation efficiency of RhB reached 93.3% and total organic carbon removal of 17.5 within 12 min. The authors also evaluated the catalyst for the degradation of other organic pollutants such as Orange acid 7 (AO7), Tetracycline hydrochloride (TCH) and Levofloxacin (LFX). The degradation efficiencies of AO7 and LFX were 97.8 and 46.2%, respectively. Based on EPR and quenching experiments (Fig. 25), radical ($SO_4^{\bullet}$, $HO^{\bullet}$) and non-radical ($^1O_2$) mechanism were involved in Co-HAP-2/PMS system. Similarly, Song et al. [149] reported a strategy for simultaneous cobalt removal and organic waste decomposition using HAP@Carbon. After Co^{2+} adsorption using HAP@C, the resulting material was used for activation of PMS to MB degradation.

4.2.2 Photocatalysis

In recent years, hydroxyapatite has been also used in the photocatalytic degradation of organic pollutants in water. However, the stability and photocatalytic activity of pure HAp remain unsatisfactory due to its insulating nature. For this reason, several studies have tried to combined HAp with other photocatalytic materials to improve its photocatalytic efficiency. In this regard, Bekkali et al. [150] reported the preparation of ZnO/HAp composites via dissolution–precipitation of natural phosphate. The

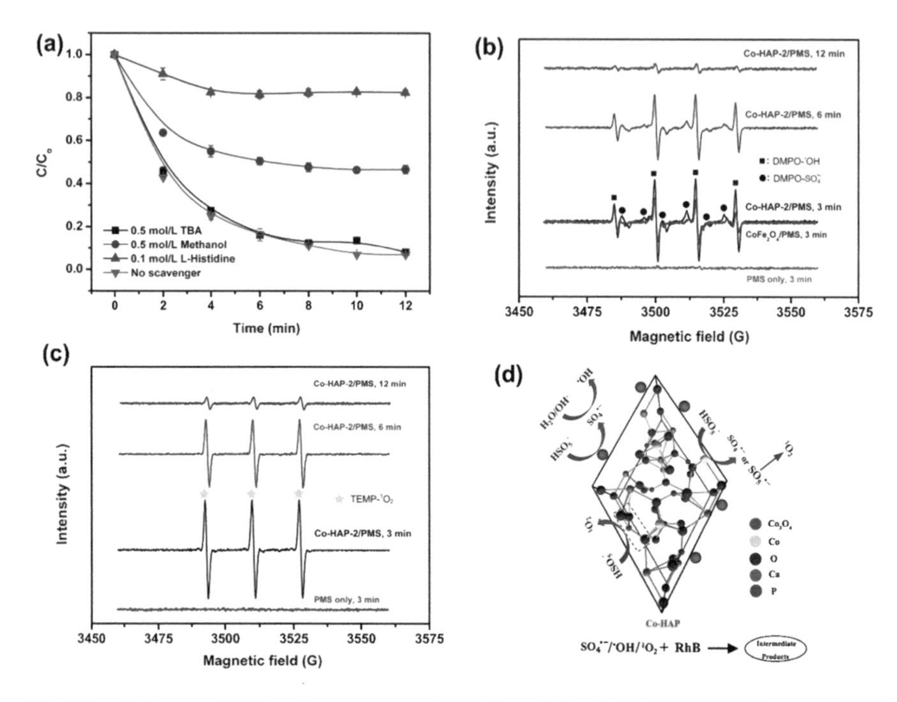

Fig. 25 **a** Influence of different scavengers on RhB degradation in Co-HAP-2/PMS system. EPR spectra of **b** DMPO and **c** TEMP in PMS activation via Co-HAP-2 catalyst. **d** The possible mechanism of ROS generation in Co-HAP-2/PMS system. Reproduced with permission from [148]. Copyright 2020 Elsevier

as-prepared composites with different amounts of ZnO were evaluated for the degradation of ciprofloxacin and ofloxacin antibiotics under exposure to UV irradiation. As mentioned above, the pure HAp does not have photocatalytic activity. However, the photocatalytic efficiency of HAp increased when it was combined with small amount of ZnO. The degradation efficiency using 25ZnO/HAP for ciprofloxacin was 100% in only 20 min (1 h min for ofloxacin). Few combinations with other photocatalysts were also reported. For example, Xu et al. [151] prepared g-C_3N_4/HAp composites via hydrothermal route and their application for the photocatalytic degradation of tetracycline under UV irradiation. The porous hollow HAp microspheres decorated with small amount of ultrathin g-C_3N_4 (1.5%) showed an excellent photocatalytic activity, which was faster than those over pristine HAp and g-C_3N_4. Importantly, g-C_3N_4/HAp being cyclable, confirming its promising potential as an efficient photocatalyst. In another report, Zou et al. [152] used hollow HAp combined with red phosphorous as photocatalyst for fast degradation of antibiotic pollutants (rifampicin, tetracycline, and levofloxacin). The as-prepared material exhibits outstanding photocatalytic activity and stability, and has universal application for degradation of each antibiotics. The radicals capture experiments using EPR analysis confirmed that the superoxide radicals ($O_2^{-\bullet}$), hydroxyl radicals ($OH^\bullet$) and hole (h^+) were the main active species in the photocatalytic degradation of the antibiotics.

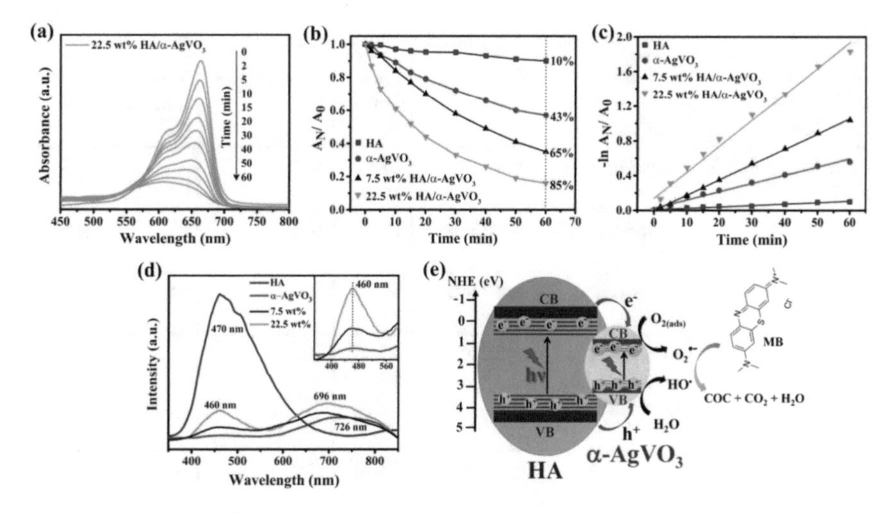

Fig. 26 **a** UV–Vis absorption of MB in solution containing 22.5 wt% HA/α-$AgVO_3$ composite, **b** discoloration efficiency, **c** first-order plot, **d** photoluminescence properties ($\lambda_{exc} = 350$ nm), and **e** photocatalytic mechanism. Reproduced with permission [153]. Copyright 2018 Elsevier

Very recently, Da Silva et al. [153] reported that the combination of α-$AgVO_3$ with HAP has led to enhanced photodegradation activity and antifungal activity. The as-prepared composites were tested for the degradation of methylene blue under UV light. The results showed that the photocatalytic performance of the α-$AgVO_3$/HAP composite was significantly enhanced compared to pure α-$AgVO_3$ or HAP under the same conditions. The degradation constant rate (k) of 22.5 wt% α-$AgVO_3$/HAP composite was 3.98 times higher than that of pristine α-$AgVO_3$. The 22.5 wt% α-$AgVO_3$/HAP composite was able to degrade 85% of MB (1.10^{-5} mol L^{-1}) in only 60 min (Fig. 26).

Table 6 summarizes the reported studies regarding the photocatalytic degradation of organic pollutants using HAp and HAp-based composites.

5 Conclusions

Owing to the advances in materials science and nanotechnology, new materials have been developed for several purposes. Among of various materials, hydroxyapatite has gained considerable attention as biomaterials for several application such adsorption, catalysis and photocatalysis. However, an in-depth understanding of the fundamental chemistry of HAp is important. This book chapter systemically summarizes the different routes of hydroxyapatite-based materials' elaboration with a display of their interesting characteristics and their use in pollutants' elimination in water. The most plausible field of application of HAp materials would be in adsorption of ions metals and fluoride in aqueous solution. Therefore, some typical examples of HAp based

Table 6 HAP and HAP based materials for photocatalytic degradation of organic pollutant

Adsorbent HAP-based materials	Pollutants, concentration (mg/L)	Type	Time (min)	Degradation (%)	TOC removal (%)	Refs.
Fe(III)-substituted HAP	Rhodamine B 5	dye	240	44.2	–	[154]
g-C_3N_4/HAP	Tetracycline, 50	Antibiotic	15	100	91	[151]
Graphene oxide/HAP	Tetracycline 60	Antibiotic	30	100	85	[155]
Red phosphorus/HAP	Rifampicin, tetracycline, levofloxacin 10	Antibiotic	10, 20, 50	100	–	[152]
ZnO/HAp	Ciprofloxacin, ofloxacin	Antibiotic	20, 60	100	–	[150]
α-$AgVO_3$ /HAP	Methylene blue	Dye	60	85	–	[153]
CdS/HAP	Tetracycline 50	Antibiotic	30	90.2	–	[156]
Fe_3O_4/HAP	Diazinon 10	insecticide	60	75	–	[157]
Ag_3PO_4/HAP	Methylene blue 20	Dye	20, 120	68, 99	–	[158]

materials and their structure–property and applications for removal of inorganic and organic contaminants in water were highlighted and discussed in detail. In addition, operating HAp as a substrate to an active phase for oil–water separation, catalyst for activation of various oxidants (H_2O_2, PMS) and as photocatalyst for degradation of organic pollutants was also discussed.

Acknowledgements The authors express their gratitude to University Hassan II, Casablanca; Moroccan Foundation for Advanced Science Innovation and Research (MAScIR); for their commitment and encouragement towards young researchers' to develop international works.

References

1. Costa DO, Dixon SJ, Rizkalla AS (2012) One- and three-dimensional growth of hydroxyapatite nanowires during sol-gel-hydrothermal synthesis. ACS Appl Mater Interfaces 4(3):1490–1499
2. Dorozhkin SV (2010) Bioceramics of calcium orthophosphates. Biomaterials 31(7):1465–1485
3. Liu M et al (2014) Multifunctional hydroxyapatite/Na(Y/Gd)F4:Yb3+, Er 3+ composite fibers for drug delivery and dual modal imaging. Langmuir 30(4):1176–1182

4. Zhang P et al (2013) Ru–Zn supported on hydroxyapatite as an effective catalyst for partial hydrogenation of benzene. Green Chem 15(1):152–159
5. Meski S, Ziani S, Khireddine H (2010) Removal of lead ions by hydroxyapatite prepared from the egg shell. J Chem Eng Data 55(9):3923–3928
6. Vila M, Sánchez-Salcedo S, Cicuéndez M, Izquierdo-Barba I, Vallet-Regí M (2011) Novel biopolymer-coated hydroxyapatite foams for removing heavy-metals from polluted water. J Hazard Mater 192(1):71–77
7. Amedlous A, Amadine O, Essamlali Y, Daanoun K, Aadil M, Zahouily M (2019) Aqueous-phase catalytic hydroxylation of phenol with H2O2 by using a copper incorporated apatite nanocatalyst. RSC Adv. 9(25):14132–14142
8. Amadine O, Essamlali Y, Amedlous A, Zahouily M (2019) Iron oxide encapsulated by copper-apatite: an efficient magnetic nanocatalyst for *N*-arylation of imidazole with boronic acid. RSC Adv. 9(62):36471–36478
9. Essamlali Y, Amadine O, Larzek M, Len C, Zahouily M (2017) Sodium modified hydroxyapatite: highly efficient and stable solid-base catalyst for biodiesel production. Energy Convers Manag 149:355–367
10. Mondal S, De Anda Reyes ME, Pal U (2017) Plasmon induced enhanced photocatalytic activity of gold loaded hydroxyapatite nanoparticles for methylene blue degradation under visible light. RSC Adv 7(14):8633–8645
11. Shariffuddin JH, Jones MI, Patterson DA (2013) Greener photocatalysts: Hydroxyapatite derived from waste mussel shells for the photocatalytic degradation of a model azo dye wastewater. Chem Eng Res Des 91(9):1693–1704
12. Liu W, Qian G, Zhang B, Liu L, Liu H (2016) Facile synthesis of spherical nano hydroxyapatite and its application in photocatalytic degradation of methyl orange dye under UV irradiation. Mater Lett 178:15–17
13. Nishikawa H (2003) Surface changes and radical formation on hydroxyapatite by UV irradiation for inducing photocatalytic activation. J Mol Catal A Chem 206(1–2):331–338
14. Kottegoda N et al (2017) Urea-hydroxyapatite nanohybrids for slow release of nitrogen. ACS Nano 11(2):1214–1221
15. Elhassani CE, Essamlali Y, Aqlil M, Nzenguet AM, Ganetri I, Zahouily M (2019) "Urea-impregnated HAP encapsulated by lignocellulosic biomass-extruded composites: A novel slow-release fertilizer. Environ Technol Innov 15:100403
16. Pang YX, Bao X (2003) Influence of temperature, ripening time and calcination on the morphology and crystallinity of hydroxyapatite nanoparticles. J Eur Ceram Soc 23(10):1697–1704
17. Wang P, Li C, Gong H, Jiang X, Wang H, Li K (2010) Effects of synthesis conditions on the morphology of hydroxyapatite nanoparticles produced by wet chemical process. Powder Technol 203(2):315–321
18. Kumar R, Prakash KH, Cheang P, Khor KA (2004) Temperature driven morphological changes of chemically precipitated hydroxyapatite nanoparticles. Langmuir 20(13):5196–5200
19. Li H, Mei L, Liu H, Liu Y, Liao L, Kumar RV (2017) Growth mechanism of surfactant-free size-controlled luminescent hydroxyapatite nanocrystallites. Cryst Growth Des 17(5):2809–2815
20. Qiao W et al (2017) Biomimetic hollow mesoporous hydroxyapatite microsphere with controlled morphology, entrapment efficiency and degradability for cancer therapy. RSC Adv 7(71):44788–44798
21. Owens GJ et al (2016) Sol-gel based materials for biomedical applications. Progress Mater Sci 77:1–79. Elsevier Ltd.
22. Sanosh KP, Chu M-C, Balakrishnan A, Kim TN, Cho S-J (2009) Preparation and characterization of nano-hydroxyapatite powder using sol-gel technique. Bull Mater Sci 32(5):465–470
23. Bigi A, Boanini E, Rubini K (2004) Hydroxyapatite gels and nanocrystals prepared through a sol–gel process. J Solid State Chem 177(9):3092–3098
24. Liu DM, Troczynski T, Tseng WJ (2001) Water-based sol-gel synthesis of hydroxyapatite: process development. Biomaterials 22(13):1721–1730

25. Ishikawa K, Garskaite E, Kareiva A Sol–gel synthesis of calcium phosphate-based biomaterials—a review of environmentally benign, simple, and effective synthesis routes. J Sol-Gel Sci Technol
26. Gross KA, Chai CS, Kannangara GSK, Ben-Nissan B, Hanley L (1998) Thin hydroxyapatite coatings via sol-gel synthesis. J Mater Sci Mater Med 9(12):839–843
27. Grigoraviciute-Puroniene I, Stankeviciute Z, Ishikawa K, Kareiva A (2020) Formation of calcium hydroxyapatite with high concentration of homogeneously distributed silver. Microporous Mesoporous Mater 293:109806
28. Brendel T, Engel A, Rüssel C (1992) Hydroxyapatite coatings by a polymeric route. J Mater Sci Mater Med 3(3):175–179
29. Amer W et al (2014) Microwave-assisted synthesis of mesoporous nano-hydroxyapatite using surfactant templates. CrystEngComm 16(4):543–549
30. Qi C et al (2013) Hydroxyapatite hierarchically nanostructured porous hollow microspheres: Rapid, sustainable microwave-hydrothermal synthesis by using creatine phosphate as an organic phosphorus source and application in drug delivery and protein adsorption. Chem A Eur J 19(17):5332–5341
31. Kay MI, Young RA, Posner AS (1964) Crystal structure of hydroxyapatite. Nature 204(4963):1050–1052
32. Ma G, Liu XY (2009) Hydroxyapatite: hexagonal or monoclinic? Cryst Growth Des 9(7):2991–2994
33. Basirun WJ, Nasiri-Tabrizi B, Baradaran S (2018) Overview of Hydroxyapatite-graphene nanoplatelets composite as bone graft substitute: mechanical behavior and in-vitro biofunctionality. Crit Rev Solid State Mater Sci 43(3):177–212
34. Stipniece L, Salma-Ancane K, Borodajenko N, Sokolova M, Jakovlevs D, Berzina-Cimdina L (2013) CERAMICS characterization of Mg-substituted hydroxyapatite synthesized by wet chemical method. Ceram Int 40:3261–3267
35. Bigi A, Boanini E, Capuccini C, Gazzano M (2006) Strontium-substituted hydroxyapatite nanocrystals
36. Yasukawa A, Higashijima M, Kandori K, Ishikawa T (2005) Preparation and characterization of cadmium-calcium hydroxyapatite solid solution particles. Physicochem Eng Asp 268:111–117
37. Suchanek WL, Byrappa K, Shuk P, Riman RE, Janas VF, Tenhuisen KS (2004) Preparation of magnesium-substituted hydroxyapatite powders by the mechanochemical-hydrothermal method. Biomaterials 25(19):4647–4657
38. Miyaji F, Kono Y, Suyama Y (2005) Formation and structure of zinc-substituted calcium hydroxyapatite. Mater Res Bull 40(2):209–220
39. Yasukawa A, Ueda E, Kandori K, Ishikawa T (2005) Preparation and characterization of carbonated barium-calcium hydroxyapatite solid solutions. J Colloid Interface Sci 288(2):468–474
40. Zhu K, Yanagisawa K, Shimanouchi R, Onda A, Kajiyoshi K (2006) Preferential occupancy of metal ions in the hydroxyapatite solid solutions synthesized by hydrothermal method. J Eur Ceram Soc 26(4–5):509–513
41. Neacsu IA, Stoica AE, Vasile BS, Andronescu E (2019) Luminescent hydroxyapatite doped with rare earth elements for biomedical applications. Nanomater (Basel, Switzerland) 9(2)
42. Silvester L et al (2014) Structural, textural and acid–base properties of carbonate-containing hydroxyapatites. J Mater Chem A 2(29):11073–11090
43. Zapanta-Legeros R (1965) Effect of carbonate on the lattice parameters of apatite. Nature 206(4982):403–404. Nature Publishing Group
44. Landi E, Celotti G, Logroscino G, Tampieri A (2003) Carbonated hydroxyapatite as bone substitute. J Eur Ceram Soc 23(15):2931–2937
45. Ramesh K, Ling EGY, Gwie CG, White TJ, Borgna A (2012) Structure and surface reactivity of WO_4^{2-}, SO_4^{2-}, PO_4^{3-} Modified Ca-hydroxyapatite catalysts and their activity in ethanol conversion. J Phys Chem C 116(35):18736–18745

46. Cacciotti I (2016) Cationic and anionic substitutions in hydroxyapatite. In: Handbook of bioceramics and biocomposites. Springer International Publishing, Cham, pp 145–211
47. Cazalbou S, Combes C, Eichert D, Rey C (2004) Adaptative physico-chemistry of bio-related calcium phosphates. J Mater Chem 14(14):2148
48. Silvester L et al (2015) Reactivity of ethanol over hydroxyapatite-based Ca-enriched catalysts with various carbonate contents. Catal Sci Technol 5(5):2994–3006
49. Sugiyama S, Osaka T, Hirata Y, Sotowa KI (2006) Enhancement of the activity for oxidative dehydrogenation of propane on calcium hydroxyapatite substituted with vanadate. Appl Catal A Gen 312(1–2):52–58
50. Sun R, Chen K, Liao Z, Meng N (2013) Controlled synthesis and thermal stability of hydroxyapatite hierarchical microstructures. Mater Res Bull 48(3):1143–1147
51. Kannan S, Ferreira JMF (2006) Synthesis and thermal stability of hydroxyapatite-β-tricalcium phosphate composites with cosubstituted sodium, magnesium, and fluorine. Chem Mater 18(1):198–203
52. Tõnsuaadu K, Gross KA, Pluduma L, Veiderma M (2012) A review on the thermal stability of calcium apatites. J Therm Anal Calorim 110(2):647–659
53. Wang T, Dorner-Reisel A, Müller E (2004) Thermogravimetric and thermokinetic investigation of the dehydroxylation of a hydroxyapatite powder. J Eur Ceram Soc 24(4):693–698
54. Trombe JC, Montel G (1978) Some features of the incorporation of oxygen in different oxidation states in the apatitic lattice—I On the existence of calcium and strontium oxyapatites. J Inorg Nucl Chem 40(1):15–21
55. Kijima T, Tsutsumi M (1979) Preparation and thermal properties of dense polycrystalline oxyhydroxyapatite. J Am Ceram Soc 62(9–10):455–460
56. Liao C-J, Lin F-H, Chen K-S, Sun J-S (1999) Thermal decomposition and reconstitution of hydroxyapatite in air atmosphere. Biomaterials 20(19):1807–1813
57. Gross KA, Berndt CC (1998) Thermal processing of hydroxyapatite for coating production. J Biomed Mater Res 39(4):580–587
58. Malina D, Biernat K, Sobczak-Kupiec A (2013) Studies on sintering process of synthetic hydroxyapatite. Acta Biochim Pol 60(4):851–855
59. Kamieniak J, Kelly PJ, Banks CE, Doyle AM (2018) Mechanical, pH and thermal stability of mesoporous hydroxyapatite. J Inorg Organomet Polym Mater 28(1):84–91
60. Bell LC, Mika H, Kruger BJ (1978) Synthetic hydroxyapatite-solubility product and stoichiometry of dissolution. Arch Oral Biol 23(5):329–336
61. Tsuchida T, Kubo J, Yoshioka T, Sakuma S, Takeguchi T, Ueda W (2008) Reaction of ethanol over hydroxyapatite affected by Ca/P ratio of catalyst. J Catal 259(2):183–189
62. Diallo-Garcia S et al (2014) Identification of surface basic sites and acid-base pairs of hydroxyapatite. J Phys Chem C 118(24):12744–12757
63. Ali I (2012) New generation adsorbents for water treatment. Chem Rev 112(10): 5073–5091. American Chemical Society
64. Hao L, Lv Y, Song H (2017) The morphological evolution of hydroxyapatite on high-efficiency Pb^{2+} removal and antibacterial activity. Microchem J 135:16–25
65. Da Rocha NCC, De Campos RC, Rossi AM, Moreira EL, Barbosa ADF, Moure GT (2002) Cadmium uptake by hydroxyapatite synthesized in different conditions and submitted to thermal treatment. Environ Sci Technol 36(7):1630–1635
66. Smičiklas I, Onjia A, Raičević S, Janaćković D, Mitrić M (2008) Factors influencing the removal of divalent cations by hydroxyapatite. J Hazard Mater 152(2):876–884
67. Campisi S, Castellano C, Gervasini A (2018) Tailoring the structural and morphological properties of hydroxyapatite materials to enhance the capture efficiency towards copper(II) and lead(II) ions. New J Chem 42(6):4520–4530
68. Guan DX, Ren C, Wang J, Zhu Y, Zhu Z, Li W (2018) Characterization of lead uptake by nano-sized hydroxyapatite: a molecular scale perspective. ACS Earth Sp. Chem. 2(6):599–607
69. Oulguidoum A, Bouyarmane H, Laghzizil A, Nunzi JM, Saoiabi A (2019) Development of sulfonate-functionalized hydroxyapatite nanoparticles for cadmium removal from aqueous solutions. Colloids Interface Sci Commun 30:100178

70. Fang X, Zhu S, Ma J, Wang F, Xu H, Xia M (2020) The facile synthesis of zoledronate functionalized hydroxyapatite amorphous hybrid nanobiomaterial and its excellent removal performance on Pb^{2+} and Cu^{2+}. J Hazard Mater 392:122291
71. Wang M et al (2019) Unexpectedly high adsorption capacity of esterified hydroxyapatite for heavy metal removal. Langmuir 35(49):16111–16119
72. Saoiabi S, El Asri S, Laghzizil A, Masse S, Ackerman JL (2012) Synthesis and characterization of nanoapatites organofunctionalized with aminotriphosphonate agents. J Solid State Chem 185:95–100
73. Saoiabi S, Gouza A, Bouyarmane H, Laghzizil A, Saoiabi A (2016) Organophosphonate-modified hydroxyapatites for Zn(II) and Pb(II) adsorption in relation of their structure and surface properties. J Environ Chem Eng 4(1):428–433
74. Oulguidoum A, Bouiahya K, Bouyarmane H, Talbaoui A, Nunzi JM, Laghzizil A (2021) Mesoporous nanocrystalline sulfonated hydroxyapatites enhance heavy metal removal and antimicrobial activity. Sep Purif Technol 255:117777
75. Ma J, Xia M, Zhu S, Wang F (2020) A new alendronate doped HAP nanomaterial for Pb^{2+}, Cu^{2+} and Cd^{2+} effect absorption. J Hazard Mater 400:123143
76. Salah TA, Mohammad AM, Hassan MA, El-Anadouli BE (2014) Development of nano-hydroxyapatite/chitosan composite for cadmium ions removal in wastewater treatment. J Taiwan Inst Chem Eng 45(4):1571–1577
77. Lei Y, Guan JJ, Chen W, Ke QF, Zhang CQ, Guo YP (2015) Fabrication of hydroxyapatite/chitosan porous materials for Pb(ii) removal from aqueous solution. RSC Adv 5(32):25462–25470
78. Na Y, Lee J, Lee SH, Kumar P, Kim JH, Patel R (2020) Removal of heavy metals by polysaccharide: a review. Polym-Plast Technol Mater 59(16): 1770–1790. Bellwether Publishing, Ltd.
79. Googerdchian F, Moheb A, Emadi R (2012) Lead sorption properties of nanohydroxyapatite-alginate composite adsorbents. Chem Eng J 200–202:471–479
80. Saber-Samandari S, Saber-Samandari S, Gazi M (2013) Cellulose-graft-polyacrylamide/hydroxyapatite composite hydrogel with possible application in removal of Cu (II) ions. React Funct Polym 73(11):1523–1530
81. Manatunga DC, De Silva RM, De Silva KMN, Ratnaweera R (2016) Natural polysaccharides leading to super adsorbent hydroxyapatite nanoparticles for the removal of heavy metals and dyes from aqueous solutions. RSC Adv 6(107):105618–105630
82. Ferri M, Campisi S, Scavini M, Evangelisti C, Carniti P, Gervasini A (2019) In-depth study of the mechanism of heavy metal trapping on the surface of hydroxyapatite. Appl Surf Sci 475:397–409
83. Ferri M, Campisi S, Gervasini A (2019) Nickel and cobalt adsorption on hydroxyapatite: a study for the de-metalation of electronic industrial wastewaters. Adsorption 25(3):649–660
84. Wijesinghe WPSL et al (2018) Preparation and characterization of mesoporous hydroxyapatite with non-cytotoxicity and heavy metal adsorption capacity. New J Chem 42(12):10271–10278
85. Su M et al (2019) Removal of U(VI) from nuclear mining effluent by porous hydroxyapatite: evaluation on characteristics, mechanisms and performance. Environ Pollut 254
86. Ansari A, Vahedi S, Tavakoli O, Khoobi M, Faramarzi MA (2019) Novel Fe_3O_4/hydroxyapatite/β-cyclodextrin nanocomposite adsorbent: synthesis and application in heavy metal removal from aqueous solution. Appl Organomet Chem 33(1):e4634
87. Zhang Q, Dan S, Du K (2017) Fabrication and characterization of magnetic hydroxyapatite entrapped agarose composite beads with high adsorption capacity for heavy metal removal. Ind Eng Chem Res 56(30):8705–8712
88. Akafu T, Chimdi A, Gomoro K (2019) Removal of fluoride from drinking water by sorption using diatomite modified with aluminum hydroxide
89. Zhou J et al (2018) Highly selective and efficient removal of fluoride from ground water by layered Al–Zr–La Tri-metal hydroxide. Appl Surf Sci 435:920–927
90. Liu Y, Fan Q, Wang S, Liu Y, Zhou A, Fan L (2016) Adsorptive removal of fluoride from aqueous solutions using Al-humic acid-La aerogel composites. Chem Eng J 306:174–185

91. Samant A, Nayak B, Misra PK (2017) Kinetics and mechanistic interpretation of fluoride removal by nanocrystalline hydroxyapatite derived from limacine artica shells. J Environ Chem Eng 5(6):5429–5438
92. Tomar G, Thareja A, Sarkar S (2015) Enhanced fluoride removal by hydroxyapatite-modified activated alumina. Int J Environ Sci Technol 12(9):2809–2818
93. Pandi K, Viswanathan N (2015) Synthesis of alginate beads filled with nanohydroxyapatite: an efficient approach for fluoride sorption. J Appl Polym Sci 132(19):n/a–n/a
94. Nayak B, Samant A, Patel R, Misra PK (2017) Comprehensive understanding of the kinetics and mechanism of fluoride removal over a potent nanocrystalline hydroxyapatite surface. ACS Omega
95. Zhang D et al (2012) Utilization of waste phosphogypsum to prepare hydroxyapatite nanoparticles and its application towards removal of fluoride from aqueous solution. J Hazard Mater 241–242:418–426
96. Singh S, Khare A, Chaudhari S (2020) Enhanced fluoride removal from drinking water using non-calcined synthetic hydroxyapatite. J Environ Chem Eng 8(2):103704
97. Muthu Prabhu S, Meenakshi S (2014) Synthesis of surface coated hydroxyapatite powders for fluoride removal from aqueous solution. Powder Technol 268(1):306–315
98. Yu X, Tong S, Ge M, Zuo J (2013) Removal of fluoride from drinking water by cellulose@hydroxyapatite nanocomposites. Carbohydr Polym 92(1):269–275
99. Sairam Sundaram C, Viswanathan N, Meenakshi S (2008) Uptake of fluoride by nano-hydroxyapatite/chitosan, a bioinorganic composite. Bioresour Technol 99(17):8226–8230
100. Sairam Sundaram C, Viswanathan N, Meenakshi S (2009) Fluoride sorption by nano-hydroxyapatite/chitin composite. J Hazard Mater 172(1):147–151
101. Norvill ZN, Shilton A, Guieysse B (2016) Emerging contaminant degradation and removal in algal wastewater treatment ponds: identifying the research gaps. J Hazard Mater 313:291–309. Elsevier B.V
102. Rojas S, Horcajada P (2020) Metal–organic frameworks for the removal of emerging organic contaminants in water. Chem Rev 120(16):8378–8415
103. Anfar Z et al (2019) Recent trends on numerical investigations of response surface methodology for pollutants adsorption onto activated carbon materials: a review. Crit Rev Environ Sci Technol 1–42
104. Anfar Z et al (2019) High extent mass recovery of alginate hydrogel beads network based on immobilized bio-sourced porous carbon@Fe_3O_4–NPs for organic pollutants uptake. Chemosphere 236:124351
105. Zbair M, Anfar Z, Ahsaine HA (2019) Reusable bentonite clay: modelling and optimization of hazardous lead and p-nitrophenol adsorption using a response surface methodology approach. RSC Adv 9(10):5756–5769
106. Chahkandi M (2017) Mechanism of Congo red adsorption on new sol-gel-derived hydroxyapatite nano-particle. Mater Chem Phys 202:340–351
107. Mehri A, Ben Moussa S, Laghzizil A, Nunzi JM, Badraoui B (2020) A new in situ enhancement of the hydroxyapatite surface by Tyramine: preparation and interfacial properties. Colloids Surfaces A Physicochem Eng Asp 592:124590
108. Wang G et al (2016) Simple combination of humic acid with biogenic hydroxyapatite achieved highly efficient removal of methylene blue from queous solution. RSC Adv 6(72):67888–67897
109. Guan Y et al (2018) A novel polyalcohol-coated hydroxyapatite for the fast adsorption of organic dyes. Colloids Surf A Physicochem Eng Asp 548:85–91
110. Cui L, Wang Y, Hu L, Gao L, Du B, Wei Q (2015) Mechanism of Pb(II) and methylene blue adsorption onto magnetic carbonate hydroxyapatite/graphene oxide. RSC Adv 5(13):9759–9770
111. Chatterjee S et al (2018) Scalable synthesis of hide substance–chitosan–hydroxyapatite: novel biocomposite from industrial wastes and its efficiency in dye removal. ACS Omega
112. Varaprasad K, Nunez D, Yallapu MM, Jayaramudu T, Elgueta E, Oyarzun P (2018) Nano-hydroxyapatite polymeric hydrogels for dye removal. RSC Adv 8(32):18118–18127

113. Idowu Adeogun A, Andrew Ofudje E, Abidemi Idowu M, Olateju Kareem S, Vahidhabanu S, Ramesh Babu B (2018) Biowaste-derived hydroxyapatite for effective removal of reactive yellow 4 dye: equilibrium, kinetic, and thermodynamic studies
114. Amadine O, Essamlali Y, Fihri A, Larzek M, Zahouily M (2017) Effect of calcination temperature on the structure and catalytic performance of copper—ceria mixed oxide ... and catalytic performance of copper— ceria mixed. RSC Adv 7(March):12586–12597
115. Narwade VN, Khairnar RS, Kokol V (2017) In-situ synthesised hydroxyapatite-loaded films based on cellulose nanofibrils for phenol removal from wastewater. Cellulose 24(11):4911–4925
116. Lin K, Pan J, Chen Y, Cheng R, Xu X (2009) Study the adsorption of phenol from aqueous solution on hydroxyapatite nanopowders. J Hazard Mater 161(1):231–240
117. Bouiahya K et al (2019) Synthesis and properties of alumina-hydroxyapatite composites from natural phosphate for phenol removal from water. Colloid Interface Sci Commun 31:100188
118. Fierascu I et al (2017) Efficient removal of phenol from aqueous solutions using hydroxyapatite and substituted hydroxyapatites. React Kinet Mech Catal 122(1):155–175
119. Wang X (2011) Preparation of magnetic hydroxyapatite and their use as recyclable adsorbent for phenol in wastewater. Clean: Soil, Air, Water 39(1):13–20
120. Bouiahya K, Oulguidoum A, Laghzizil A (2020) Alumina-hydroxyapatite nanocomposites and their applications for the removal of phenolic compounds from water: a comparative study. In: E3S web of conferences, vol 150, p 02008
121. Khallok H et al (2021) Ceramic hydroxyapatite foam as a new material for Bisphenol A removal from contaminated water. Environ Sci Pollut Res 1–13
122. Puga A, Rosales E, Sanromán MA, Pazos M (2020) Environmental application of monolithic carbonaceous aerogels for the removal of emerging pollutants. Chemosphere 248:125995
123. Li Y, Wang S, Zhang Y, Han R, Wei W (2017) Enhanced tetracycline adsorption onto hydroxyapatite by Fe(III) incorporation. J Mol Liq 247:171–181
124. Li Z, Li M, Wang Z, Liu X (2020) Coadsorption of Cu(II) and tylosin/sulfamethoxazole on biochar stabilized by nano-hydroxyapatite in aqueous environment. Chem Eng J 381:122785
125. Yuan L et al (2019) Influences of pH and metal ions on the interactions of oxytetracycline onto nano-hydroxyapatite and their co-adsorption behavior in aqueous solution. J Colloid Interface Sci 541:101–113
126. Ersan M, Guler UA, Acikel U, Sarioglu M (2015) Synthesis of hydroxyapatite/clay and hydroxyapatite/pumice composites for tetracycline removal from aqueous solutions. Process Saf Environ Prot 96:22–32
127. Huang B, Xiong D, Zhao T, He H, Pan X (2016) Adsorptive removal of PPCPs by biomorphic HAP templated from cotton. Water Sci Technol 74(1):276–286
128. Sharma P, Rohilla D, Chaudhary S, Kumar R, Singh AN (2019) Nanosorbent of hydroxyapatite for atrazine: A new approach for combating agricultural runoffs. Sci Total Environ 653:264–273
129. Schrope M (2011) Oil spill: deep wounds. Nature 472(7342):152–154. Nature Publishing Group
130. Wang B, Liang W, Guo Z, Liu W (2015) Biomimetic super-lyophobic and super-lyophilic materials applied for oil/water separation: a new strategy beyond nature. Chem Soc Rev 44(1):336–361 Royal Society of Chemistry
131. Ge J, Zhao HY, Zhu HW, Huang J, Shi LA, Yu SH (2016) Advanced sorbents for oil-spill cleanup: recent advances and future perspectives. Adv Mater 28(47):10459–10490. Wiley-VCH Verlag
132. Padaki M et al (2015) Membrane technology enhancement in oil-water separation. A review. Desalination 357:197–207. Elsevier
133. Zhu Y, Wang D, Jiang L, Jin J (2014) Recent progress in developing advanced membranes for emulsified oil/water separation. NPG Asia Mater 6(5):101. Nature Publishing Group
134. Ivshina IB et al (2015) Oil spill problems and sustainable response strategies through new technologies. Environ Sci Process Impacts 17(7):1201–1219. Royal Society of Chemistry

135. Xue Z, Cao Y, Liu N, Feng L, Jiang L (2014) Special wettable materials for oil/water separation. J Mater Chem A 2(8):2445–2460. The Royal Society of Chemistry
136. Li Z et al (2018) Superhydrophobic/superoleophilic polycarbonate/carbon nanotubes porous monolith for selective oil adsorption from water. ACS Sustain. Chem. Eng. 6(11):13747–13755
137. Dunderdale GJ, Urata C, Sato T, England MW, Hozumi A (2015) Continuous, high-speed, and efficient oil/water separation using meshes with antagonistic wetting properties. ACS Appl Mater Interfaces 7(34):18915–18919
138. Wang G et al (2015) Low drag porous ship with superhydrophobic and superoleophilic surface for oil spills cleanup. ACS Appl Mater Interfaces 7(47):26184–26194
139. Chen FF, Zhu YJ, Chen F, Dong LY, Yang RL, Xiong ZC (2018) Fire alarm wallpaper based on fire-resistant hydroxyapatite nanowire inorganic paper and graphene oxide thermosensitive sensor. ACS Nano 12(4):3159–3171
140. Lu BQ, Zhu YJ, Chen F (2014) Highly flexible and nonflammable inorganic hydroxyapatite paper. Chem A Eur J 20(5):1242–1246
141. Chen FF, Zhu YJ, Xiong ZC, Sun TW, Shen YQ (2016) Highly flexible superhydrophobic and fire-resistant layered inorganic paper. ACS Appl Mater Interfaces 8(50):34715–34724
142. Yang RL, Zhu YJ, Chen FF, Dong LY, Xiong ZC (2017) Luminescent, fire-resistant, and water-proof ultralong hydroxyapatite nanowire-based paper for multimode anticounterfeiting applications. ACS Appl Mater Interfaces 9(30):25455–25464
143. Yang RL, Zhu YJ, Chen FF, Qin DD, Xiong ZC (2018) Recyclable, fire-resistant, superhydrophobic, and magnetic paper based on ultralong hydroxyapatite nanowires for continuous oil/water separation and oil collection. ACS Sustain Chem Eng 6(8):10140–10150
144. Li Q, Wu D, Huang J, Guo Z (2019) Kevlar fiber-reinforced multifunctional superhydrophobic paper for oil-water separation and liquid transportation. New J Chem 43(38):15453–15461
145. Zhang YG, Zhu YJ, Xiong ZC, Wu J, Chen F (2018) Bioinspired ultralight inorganic aerogel for highly efficient air filtration and oil-water separation. ACS Appl Mater Interfaces 10(15):13019–13027
146. Das KC, Das B, Dhar SS (2020) Effective catalytic degradation of organic dyes by nickel supported on hydroxyapatite-encapsulated cobalt ferrite (Ni/HAP/$CoFe_2O_4$) magnetic novel nanocomposite. Water Air Soil Pollut 231(2):43
147. Das KC, Dhar SS (2020) Remarkable catalytic degradation of malachite green by zinc supported on hydroxyapatite encapsulated magnesium ferrite (Zn/HAP/MgFe2O4) magnetic novel nanocomposite. J Mater Sci 55(11):4592–4606
148. Pang Y, Kong L, Chen D, Yuvaraja G, Mehmood S (2020) Facilely synthesized cobalt doped hydroxyapatite as hydroxyl promoted peroxymonosulfate activator for degradation of Rhodamine B. J Hazard Mater 384:121447
149. Song F, Zhang H, Wang S, Liu L, Tan X, Liu S (2018) Atomic-level design of CoOH+-hydroxyapatite@C catalysts for superfast degradation of organics: via peroxymonosulfate activation. Chem Commun 54(39):4919–4922
150. El Bekkali C et al (2018) Zinc oxide-hydroxyapatite nanocomposite photocatalysts for the degradation of ciprofloxacin and ofloxacin antibiotics. Colloids Surf Physicochem Eng Asp 539:364–370
151. Xu T et al (2019) New and stable g-C3N4/HAp composites as highly efficient photocatalysts for tetracycline fast degradation. Appl Catal B Environ 245:662–671
152. Zou R, Xu T, Lei X, Wu Q, Xue S (2020) Novel and efficient red phosphorus/hollow hydroxyapatite microsphere photocatalyst for fast removal of antibiotic pollutants. J Phys Chem Solids 139:109353
153. da Silva et al JS (2020) Enhanced photocatalytic and antifungal activity of hydroxyapatite/α-$AgVO_3$ composites. Mater Chem Phys 252:123294
154. Liu X, Ma J, Yang J (2014) Visible-light-driven amorphous Fe(III)-substituted hydroxyapatite photocatalyst: Characterization and photocatalytic activity. Mater Lett 137:256–259
155. Zou R, Xu T, Lei X, Wu Q, Xue S (2020) Novel design of porous hollow hydroxyapatite microspheres decorated by reduced graphene oxides with superior photocatalytic performance for tetracycline removal. Solid State Sci 99:106067

156. Lei X, Xu T, Yao W, Wu Q, Zou R (2020) Hollow hydroxyapatite microspheres modified by CdS nanoparticles for efficiently photocatalytic degradation of tetracycline. J Taiwan Inst Chem Eng 106:148–158
157. Yang Z, Gong X, Zhang C (2010) Recyclable Fe3O4/hydroxyapatite composite nanoparticles for photocatalytic applications. Chem Eng J 165(1):117–121
158. Hong X et al (2012) Hydroxyapatite supported Ag 3 PO 4 nanoparticles with higher visible light photocatalytic activity. Appl Surf Sci 258(10):4801–4805

Sequestration of Heavy Metal Pollutants by Fe_3O_4-based Composites

Linda Ouma and Martin Onani

Abstract Heavy metal pollution poses a grave environmental threat. Some of the most toxic metals are highly mobile and, therefore, easily transported through ground water systems, thus, affecting large areas. Over the last decade, adsorption has been greatly focused on as a strategy for contaminated water treatment. Its versatility and relative ease of application have been a major determinant of its preference. Nano-sized adsorbents have high surface areas and are size tunable and, hence, have been favored in adsorption applications. The magnetic properties of nanosized magnetite (Fe_3O_4) have made them particularly favorable. Magnetite composites with various materials have widely been applied in the adsorptive treatment of real and synthetic water containing heavy metal pollutants. This review outlines the application of Fe_3O_4 nanoparticles and Fe_3O_4 organic composites in the adsorption of heavy metal ions in aqueous solution. The reviewed articles indicate that the formation of Fe_3O_4 inorganic–organic composites improves the adsorption efficiencies of the composites and improves their applicability by providing magnetic separability. The presence of Fe_3O_4 nanoparticles in the composite materials also provides for improved reusability of the adsorbent. Generally, the formation of these composites tends to make adsorption a more viable alternative to conventional water treatment options for heavy metal pollutants in water.

Keywords Magnetite · Composites · Adsorption efficiency · Heavy metals · Magnetism · Reusability

---

L. Ouma · M. Onani
Department of Chemistry, University of the Western Cape, Private Bag X17, Bellville 7535, South Africa
e-mail: monani@uwc.ac.za

L. Ouma (✉)
Department of Science, Technology and Engineering, Kibabii University, P. O. Box 1699, Bungoma 50200, Kenya

E. Lichtfouse et al. (eds.), *Inorganic-Organic Composites for Water and Wastewater Treatment*, Environmental Footprints and Eco-design of Products and Processes,
https://doi.org/10.1007/978-981-16-5916-4_4

1 Introduction

The environmental accumulation of heavy metals is of great concern owing to their non-biodegradability [5, 31, 64]. Heavy metal pollution occurs primarily through either of the following anthropogenic processes: manufacturing, mining, burning of fossil fuels, and agriculture [20, 66]. Although anthropogenic activities contribute the greater extent of heavy metal pollution, natural phenomena, e.g., erosion and weathering of rocks also contribute to the pollution burden [48]. According to the US EPA, the most toxic heavy metals are arsenic and lead with a maximum contaminant level goal (MCLG) of 0 mg L^{-1} (US EPA 2009; [21, 63]. Other listed toxic heavy metals are copper, chromium, mercury, nickel, and cadmium. Heavy metals may be toxic even at low concentrations resulting in poisoning or genetic disorders as they have the potential to interfere with biological processes [12, 22]. As information on the toxicity of heavy metals increases, the regulatory limits are adjusted to lower concentrations making remediation more challenging [62]. Techniques like electrochemical and photocatalytic oxidation, chemical coagulation, ion exchange, bio- and phyto-remediation, and adsorption have been employed for the adsorption of heavy metal pollution control [9, 40].

Adsorption is considered favorable due to its efficiency, versatility, simplicity of operation, zero sludge production, and relatively lower costs [42, 70]. Adsorption at the solid-solution interface provides a possibility to control pollution due to liquid waste [20]. Through consistent improvement efforts, several adsorbents have been developed with current technologies focusing on nanosized adsorbents due to the uniqueness of the properties owing to their nanometer sizes. Some of the most investigated nanomaterials are iron oxides as a result of their stability, pollutant affinity, and relatively low toxicity compared to other metal containing nanoparticles [61]. Magnetite has received great consideration because it offers superior advantages such as surface areas >100 m^2 g^{-1} and superparamagnetism (~90 emu g^{-1} for bulk magnetite) as the size reduces to nanoscale [27, 61]. The removal of pollutants through adsorption methods is highly dependent on the adsorbent's surface charge and the adsorbate's speciation and degree of ionization [20]. The presence of both ferrous (Fe^{2+}) and ferric (Fe^{3+}) ions allows Fe_3O_4 nanoparticles to participate in redox-coupled adsorption processes which are particularly useful in the sequestration of multi-valent ions. The magnetic properties of Fe_3O_4 make them easily recoverable after treatment, a challenge while using many nanometer sized materials [4, 20]. The recovered particles can be reused, therefore, reducing the economic burden of the treatment process [4, 12].

2 Magnetite

Iron-based nanoparticles have recently been applied in the adsorptive treatment of polluted water [65]. Of the reported iron-based nanoparticles, zero-valent iron has received the greatest attention [26, 32]. Nanosized iron oxides composition varies depending the iron species present and the magnetic properties; of the known iron oxides, hematite (α-Fe_2O_3) maghemite (γ-Fe_2O_3), and magnetite (Fe_3O_4) have been considered in the adsorption of heavy metals [29, 35, 38, 57]. Superparamagnetic iron oxides (magnetite; Fe_3O_4) are commonly applied because of the ease of post-adsorption retrieval using an external magnetic field. Upon removal of the magnetic field, the particles are demagnetized since they do not possess residual magnetization [25, 48].

A wide range of synthetic methods including solvothermal [29, 37], laser co-vaporization [54], sol–gel [23, 51], thermal decomposition [2, 52], and chemical co-precipitation [20, 43] has been used in the production of Fe_3O_4 nanoparticles. Chemical co-precipitation the most favored method because it is simple, efficient, and relatively cheaper than the above-mentioned methods [1, 68]. Chemical co-precipitation of Fe_3O_4 takes place in alkaline media, and the formation of Fe_3O_4 follows the reaction steps outlined in Eqs. 1–4 below [68].

$$Fe^{3+} + 3OH \rightarrow Fe(OH)_3 \tag{1}$$

$$Fe(OH)_3 \rightarrow FeOOH + H_2O \tag{2}$$

$$Fe^{2+} + 2OH^- \rightarrow Fe(OH)_2 \tag{3}$$

$$2FeOOH + Fe(OH)_2 \rightarrow Fe_3O_4 + 2H_2O \tag{4}$$

Apart from magnetite nanoparticles synthesized at the point of application, commercial magnetite nanoparticles are readily available and have also been applied in heavy metal adsorption. Iconaru et al. [20] synthesized 14 nm magnetite nanoparticles and compared their properties with those of commercial magnetite of 90 nm average diameters [20]. The surface area ratio of the commercial to synthesized magnetite was 7%; however, the synthesized sample showed lower crystallinity [15, 28]. When applied in the adsorption of As(V) and Cu(II), it was evident that the as-synthesized smaller particles provided better adsorption efficiencies for both species [20, 36]. The results obtained from As(V) and Cu(II) adsorption on both nanoparticle batches were modeled following a theoretical calculation of the packing density. Data from adsorption on commercial nanoparticles provided a better accuracy than synthesized sample, while As(V) data had 50% higher accuracy than Cu(II) adsorption data. The results pointed to more uniform distribution of commercial nanoparticles as compared to synthesized nanoparticles with a higher affinity for As(V) than Cu(II) resulting from differences in complexation energies in the adsorption process [19].

Further, Kumari et al. [29] studied Cr(VI) and Pb(II) adsorption on mesoporous Fe_3O_4 nanospheres synthesized using a solvothermal method [29]. Hollow nanospheres consist of a shell-like morphology of nanoparticles with a hollow core providing low densities. The hollow nanospheres were synthesized using a solvothermal method. In the solvothermal method, the solvent acts as a reducing medium reducing a small amount of the Fe^{3+} precursor to Fe^{2+}. The structure directing salt initiates nucleation to form spheres in the presence of the surfactant with the solvent controlling the size of the spheres. Ostwald ripening results in small inner spheres forming larger ones on the outer side increasing the size of the inner cavities. This results in the formation of a hollow interior with larger nanocrystals forming the outer surface. The particle diameters were determined to be 31 nm with surface areas of 11 $m^2\ g^{-1}$. Adsorption of Cr(VI) and Pb(II) ions resulted in modifications on the adsorbent surface of the with the initially rough surface appearing smooth in post-adsorption analyzes.

Luther et al. [34] synthesized Fe_3O_4 nanoparticles and studied the effects of pH and interfering anions on As(III) and As(V) adsorption [34]. The synthesized Fe_3O_4 nanomaterials had diameters of 17 nm, and the optimum pH used for adsorption studies was pH 6 since it was within the optimum range for both As species. The As(III) adsorption capacity was consistently higher than As(V) capacity after 1 h and 24 h contact time; however, a decreased binding capacity with increased contact time was observed and attributed to redox dissolution. Interference studies indicated that the presence of SO_4^{2-} affected the binding of As(III) decreasing it by up to 50% at concentrations greater than 1000 ppm, while As(V) binding of was completely eliminated at similar concentrations. The presence of PO_4^{3-} had insignificant effects on the adsorption capacity of either As species, while the presence of CO_3^{2-} decreased As(III) and As(V) binding of by up to 15% and 50%, respectively. From the highlighted studies, Fe_3O_4 has been portrayed as an efficient adsorbent for the sequestration of heavy metal ions in water. The particle size, pH, and competing ions have been identified as important factors influencing the adsorption process. Table 1 summarizes the efficiency of magnetite adsorbents in the sequestration of heavy metals.

Table 1 Application of magnetite nanoparticles for heavy metal (HM) adsorption

Preparation method	Particle size (nm)	Target pollutant	Adsorption capacity ($mg\ g^{-1}$)	References
Commercial	89.4 ± 0.6	As(V)	39.26	[20]
		Cu(II)	9.06	[20]
Co-precipitation	14.2 ± 0.3	As(V)	66.53	[20]
		Cu(II)	10.67	[20]
Co-precipitation	7.2 ± 1	Cr(VI)	13.51	[42]
Precipitation	25 ± 3	As(V)	9.72	[11]
Solvothermal	31.2	Pb(II)	11.89	[29]
		Cr(VI)	6.55	[29]
Co-precipitation	16.5 ± 0.5	As(III)	5.68	[34]
Co-precipitation	16.5 ± 0.5	As(V)	4.78	[34]

3 Magnetite Composites

Pristine Fe_3O_4 nanoparticles commonly face challenges of oxidation during preparation, handling, and adsorption resulting in changes in their dispersion and magnetic properties [46]. Similarly, the achievement of size control during Fe_3O_4 synthesis presents a challenge due to agglomeration resulting from high surface energies resulting in broad particle size distribution, insufficient dispersion, and difficulty in mass production. One of the most studied methods to control Fe_3O_4 properties during synthesis is the formation of composite materials, and composites retain the properties of both materials, therefore, providing a more versatile adsorbent. Fe_3O_4 inorganic–organic composite adsorbents are favored over pristine Fe_3O_4 as they incorporate the high surface areas, mechanical strength, and magnetism of the inorganic Fe_3O_4 component and provide functional groups from the organic material [43]. The organic functional groups provide multiple advantages of anchoring the Fe_3O_4 surfaces, surface passivation, as well as sequestration of various pollutants including heavy metals [14, 42]. In this section, inorganic–organic composites of Fe_3O_4 with some selected organic materials are reviewed with a focus on their application in heavy metal adsorption.

3.1 Magnetite-polymer Composites

The modification of Fe_3O_4 nanoparticle surfaces with organic ligands presents an avenue for both surface passivation and functionalization allowing for the targeted adsorption of desired pollutants [61]. Organic ligands control particle growth resulting in smaller particles, hence, large accessible surface areas, therefore, improving the adsorption capacities [16]. Zarnegar and Safari [68] studied polymer stabilization effects on Fe_3O_4 nanoparticle properties. They prepared Fe_3O_4 composite materials with polyethylene glycol (PEG) and polycitric acid (PCA) [68]. The synthesis was carried out in two stages; firstly, PCA-PEG-PCA copolymer macromolecules were prepared followed by the co-precipitation of ferric and ferrous ions in the presence of the copolymers. During the co-precipitation, ferric and ferrous salts were first stirred with the polymers resulting in the formation of a complex structure with surface carboxylic acid groups. Upon the addition of a base, the carboxylic acid groups promoted nucleation, while the copolymers controlled the nanoparticles growth thereby providing size control and resulting in the formation of particles of 5–10 nm. The dendritic nature of the macromolecules provided repulsion aiding in particle dispersion providing uniformly dispersed particles. The polymer-coated particles were spherical and monodisperse with 5–10 nm diameters and 66.54 emu g^{-1} saturation magnetization compared to 15–30 nm and 62.76 emu g^{-1}, respectively, for uncoated Fe_3O_4. Polymer stabilization improved the size distribution and magnetic properties of Fe_3O_4 as a result of improved crystallinity of the smaller nanoparticles [68].

Guan and co-workers prepared a core–shell nano-adsorbent consisting of a nano-magnetite core and a polyacrylic acid shell for the adsorption of Cr(III) ions from tannery effluent. A silane coupling agent aided the grafting of polyacrylic acid onto the surface of the magnetite nanoparticles. The synthesized composite material had a core size of 21 ± 5 nm and specific surface areas of 41.4 ± 0.6 m^2 g^{-1}. The saturation magnetization decreased in the order pristine Fe_3O_4 > silane/Fe_3O_4 > polyacrylic acid/silane/Fe_3O_4. The decrease is resulted from the encapsulation of the Fe_3O_4 in a polymeric shell; however, the resulting composite retained sufficient magnetism to facilitate magnetic separation within 5 min of adsorption completion. Chromium(III) adsorption was most favorable at pH 6 resulting in a percentage removal of 92.5%. The results indicated that Cr(III) ions were coordinated with the carboxyl groups on the polyacrylic acid shell.

Bhaumik et al. [8] reported on the synthesis of polypyrolle-magnetite (PPY/Fe_3O_4) nanocomposite for Cr(VI) adsorption [8]. The composite synthesis was carried out in situ through chemical oxidative polymerization [7]. Fe_3O_4 nanoparticles were spherical but appeared aggregated, but after polymerization with polypyrolle, the particles were spherical with larger particle sizes resulting from polypyrolle encapsulation of the particles. The nanocomposite presented superior adsorption properties compared to its constituents in the order PPY/ Fe_3O_4 > PPY > Fe_3O_4. Adsorption of Cr(VI) on the nanocomposite was determined to be through ion exchange and reduction [44]. The appearance of Cr(III) species on the spent adsorbent surface indicated that a portion of the bound Cr(VI) ions was reduced by the electron-rich polypyrolle groups in the composite material. The adsorbent was tested for reusability, and two cycles were deemed optimum with a 17% reduction in capacity observed in the third cycle.

Burks et al. [10] studied the characterization and chromium adsorption properties of mercaptopropionic acid-coated magnetite nanoparticles. Calculations from TGA measurements indicated that the coverage of mercaptopropionic acid on SPION surface was approximately 2.5 μmol m^{-2} [10], while FTIR results revealed that mercaptopropionic acid formed surface bonds with the SPION using the carboxylate end leaving the thiol group exposed [41]. Bands attributed to sulfonate groups indicated oxidation of the thiol groups during air drying. From the isotherm fitting, the obtained data pointed to a multilayer adsorption on a heterogenous surface. At low Cr(VI) concentrations, the reaction was controlled by diffusion to the adsorbent surface; however, as concentrations increased, chemisorption was the rate limiting step. Multiple rate controlling steps were confirmed by a plot of q_t against $t^{1/2}$ (intraparticle diffusion kinetic model) [43]. The adsorption mechanism was illustrated to be via the bonding of $HCrO_4^-$ ions to -SO_3H groups on the 3-MPA surface.

Alqadami et al. [1] studied the application of 5–10 nm Fe_3O_4@TSC (magnetite@tri-sodium citrate) nanocomposite in the adsorption of Cr^{3+} and Co^{2+} ions [1]. The presence of Cr–O and Co–O bonds on the spent adsorbent surface was attributed to electrostatic attraction to the electron rich acetate groups. Adsorption of Cr^{3+} was faster than that of Co^{2+}; thus, the equilibrium time for Co^{2+} was considered as the optimal contact time, and pH 6 was considered as optimal above which the formation of metal hydroxides resulted in decreased adsorption efficiency.

Langmuir isotherm and pseudo-second-order kinetics model accurately described >97% of the observed results, and the adsorption process was determined to be exothermic. A decrease in adsorption with temperature was attributed to weakening adsorbent-adsorbate and adsorbate–adsorbate forces.

A ternary composite of magnetite nanoparticles (Fe_3O_4 NPs), reduced graphene oxide sheets (rGO), and poly-*N*-phenylglycine nanofibers (d-PPG NFs) was prepared for Cu(II) adsorption [27]. The formation of Fe_3O_4 (270 $\pm$ 30 nm) on GO sheets opened the spaces between the sheets, while the grafting of PPG NFs nearly doubled the composite's surface area. The nanofibers ultrafine morphology was responsible for the increased surface area. Copper adsorption was more efficient on the ternary composite as compared to the binary composite as a result of increased affinity by PPG nanofibers and higher surface areas. The COO^- group in the nanofibers was responsible for the increased cation affinity by electrostatic attraction. Formation of a stable copper-carboxylate complex led to preferential copper adsorption in bimetal solutions with cobalt ions.

In 2010, Warner and co-workers demonstrated the synthesis of lauric acid capped Fe_3O_4 followed by a single step ligand exchange reaction to alter the surface and produce nanoparticles with affinities for a variety of heavy metal pollutants [61]. High-temperature decomposition was applied to generate a magnetite core and lauric acid shell resulting in the formation of 8 nm particles with surface areas >100 $m^2\ g^{-1}$. Surface-modified nanoparticles were applied in the adsorption of Hg, Pb, Cd, Ag, Co, Cu, and Tl in spiked river water to determine their efficiency. After ligand exchange, core sizes remained unaffected and the particles were superparamagnetic with no remnant coercivity. Adsorption efficiencies of the functionalized particles for the tested metal pollutants were consistently higher than those of activated carbon with the exception of Ag where activated carbon had the highest distribution coefficient.

Studies on organic ligand stabilized Fe_3O_4 nanoparticles have concluded that their presence does not alter the nanoparticles magnetic properties and in fact increases the particles affinity for specific heavy metal pollutants while maintaining the high surface areas and superparamagnetism [27, 68].

3.2 Magnetite-biosorbent Composites

Biological materials with the capability of binding pollutants on their surfaces (adsorption) are referred to as biosorbents. In the process of biosorption, heavy metals (pollutants) are adsorbed through a metabolically passive process which occurs on non-living tissues [67]. Biomaterials do not pose a threat to the environment since they are organic in nature and are biodegradable [47]. Several biosorbents have been applied in heavy metals adsorption due to the abundance of functional groups capable of heavy metal sequestration [13, 50, 60]. Despite the adsorption potentials of biomaterials, they face challenges such as low porosity, surface areas, and difficulty in post-treatment separation [39, 66]. The incorporation of nanomaterials on the surfaces of biosorbents has been confirmed to improve surface areas and porosity of adsorbents [27, 69]. Fe_3O_4 nanoparticles when deposited on biosorbents incorporate magnetic properties on the composite adsorbent allowing for the application of magnetic separation. In this section, we review the application of Fe_3O_4-biosorbent composites in the adsorption of heavy metals.

3.2.1 Magnetite-chitosan Composites

Chitosan is the second most naturally available polymer after cellulose, and it contains $-NH_2$ and OH functional groups which sequester ions through coordination forming a mesh-like cage-shaped structure [18, 64]. However, the reusability of traditional chitosan adsorbents poses a challenge; therefore, the formation of magnetic composites has been considered. The chitosan-magnetite composites faced some challenges due to low sorption capacities owing to their large sizes leading [64] to explore the formation of polyethylene modified polystyrene/Fe_3O_4/chitosan (PS/Fe_3O_4/CS-PEI) of sub-micron sizes for Cu(II) adsorption [64]. The adsorbents had an average size of 300 nm with Fe_3O_4 nanoparticles of ~10 nm immobilized on the surface. The composite retained its magnetic properties and was easily recovered by magnetic separation, and it was confirmed that all the constituents of the composite material were present in the adsorbent. The mechanism of Cu(II) adsorption on PS/Fe_3O_4/CS-PEI was attributed to the surface complexation between Cu(II) ions and N atoms from nitrogen containing groups on the adsorbent surface.

Haldorai et al. [18] demonstrated the efficiency of <30 nm Fe_3O_4/chitosan (Fe_3O_4/CS) for the adsorption of Lanthanum (La^{3+}) ions from aqueous solutions [18]. Successful adsorption of La^{3+} on the adsorbent surface was confirmed by scanning electron microscopy. Response surface methodology (RSM) was applied to optimize the factors affecting the adsorption process. The Box-Behnken model (BBM) was used to determine the parameters' effects on the adsorption efficiency. The investigated parameters were solution pH, adsorbent dosage, reaction time, and temperature. The quadratic model which explained 87% of the total variables predicted the efficient removal of La^{3+} for the studied parameters. The adsorption efficiency was highly dependent on the solution pH, and the optimum pH was observed to be pH 11. Reaction time and temperature had insignificant effects on La^{3+} adsorption efficiency. Increasing the adsorbent dosage provided more adsorption sites thereby increasing the adsorption efficiency. The Freundlich isotherm model fitted the adsorption data pointing to adsorption on heterogenous sites.

Chitosan-modified biochar was employed for the adsorption of dissolved As(V) by [33] to improve the separation ability of the chitosan/biochar composite, and chitosan was coated with magnetic Fe_3O_4 fluid during the composite formation [33]. Although the synthesized magnetic chitosan biochar (MCB) exhibited a lower saturation magnetization (16.67 emu g^{-1}) compared to the magnetic fluid (67 emu g^{-1}), it was sufficient to provide magnetic separation. The As(V) adsorption capacity of the binary and ternary composites improved threefold compared to biochar indicating the contribution of chitosan and Fe_3O_4 during adsorption. In the presence of competing anions, As(V) adsorption efficiency was significantly altered by the presence of PO_4^{3-}, CO_3^{2-}, and SO_4^{2-}, while Cl and NO_3^- had no significant impact.

3.2.2 Magnetite-agricultural Biosorbent Composites

Plant tannin is a natural polyphenol capable of reductively adsorbing heavy metal ions, including Ag(I), Au(III), Cr(VI), and Pd(II), due to the large number of hydroxyl groups it contains [14]. Microspheres consisting of a magnetic Fe_3O_4 core and silica shell are favorable as the magnetic core provides for simple magnetic retrieval, while the silica shell passivates the core and provides active sites allowing for further modification. Persimmon tannin (PT) was immobilized on the $Fe_3O_4@SiO_2$ spheres to create an organic–inorganic composite material and applied in the sequestration of Au(III) and Pd(II) [14]. The PT was immobilized onto the spheres via a two-step method involving the reduction of $FeCl_3$ in ethylene glycol to form Fe_3O_4 and sol–gel method to prepare the silica coating [14]. Solution pH between 1 and 5 was investigated for Au(III) and Pd(II) adsorption. There was an observed increase in Au(III) adsorption with an increase in pH which was attributed to the more favorable adsorption of hydrolyzed chlorogold ($AuCl_3(OH)^-$ and $AuCl_2(OH)^{2-}$) as compared to $AuCl_4^-$ which is the dominant species below pH 3. The decreased adsorption below the point of zero charge (pH_{PZC}) at pH 1.6 resulted from competition for the available with Cl^- ions in solution. The optimum adsorption of Pd(II) was determined to be pH 3 despite the observed slight increase in adsorption capacity at pH 5 which was attributed to the formation of $Pd(OH)_4^{2-}$ whose adsorption is less favorable than that of $PdCl_3$. The transfer and sharing electrons between the $Fe_3O_4@SiO_2@PT$ and metal ions were determined to be the mechanism for adsorption. Au(III) and Pd(II) adsorption onto $Fe_3O_4@SiO_2@PT$ proceeded via a fast adsorption phase with electrostatic adsorption and intraparticle diffusion controlling the process followed by a slower second phase resulting forms the relatively time-consuming redox process. Evidence of the redox process was obtained from the post-adsorption XPS analysis, and the spectra indicated that Au(III) was reduced to metallic gold, while Pd(II) was chelated by oxygen-containing surface groups of the adsorbent. Au(III) adsorption was overall faster than Pd(II) adsorption indicating a higher affinity of the adsorbent for Au(III). The $Fe_3O_4@SiO_2@PT$ composite demonstrated selective adsorption for Au(III) despite interference from other metal ions, while the selectivity for Pd(II) was lower due to competition for adsorption sites with Au(III). Higher concentrations of Cl^- ions also decreased Pd(II) adsorption efficiency.

Magnetite-tea waste composite was prepared by [66] for the adsorption of Pb^{2+} from rainwater, groundwater, and freshwater [66]. Tea leaves contain numerous polar aliphatic and aromatic functional groups allowing it to be good adsorbent for heavy metals [55]. The magnetite-tea waste composite was prepared via co-precipitation of iron-loaded tea waste in aqueous media resulting in a sixfold increase in the tea waste surface area with a slight reduction in pore size. The prepared composite retained the superparamagnetism of Fe_3O_4 with saturation magnetism values of 7 and 32 emu g^{-1} for the composite and Fe_3O_4, respectively. Formation of the composite prevented

Fe leaching in the studied water samples. Unmodified tea waste showed consistent higher Pb^{2+} adsorption efficiencies which is largely attributed to the presence of -NH_2 and -COOH functional groups which sequester Pb^{2+} ions, while the presence of humic acid resulted in the formation of Pb-humate complexes, therefore, lowering Pb^{2+} concentration in groundwater samples.

The calcination effects of Fe_3O_4—honeycomb briquette cinders (HBC) — composite on arsenic (As(III) and As(V)) adsorption was studied by [6]. HBC are waste biomass materials from cylindrical stoves. Arsenic adsorption on the Fe_3O_4—HBC—composite surface proceeded via a ligand exchange process and formed inner-sphere complexes [6, 45]. Electrostatic repulsion led to decreased adsorption at higher pH ranges since the adsorbent surface became increasingly negatively charged.

3.2.3 Magnetite-cellulose Composites

Cellulose is a renewable, biodegradable, and inexpensive raw material as a result of its abundance in nature; in fact, it has often been cited as the most abundant organic raw material on the planet [56]. The challenge cellulose-based adsorbents face is difficulty in recovery, and magnetization of the cellulose adsorbents through the formation of composites with superparamagnetic magnetite nanoparticles, therefore, provides a simple solution to this challenge. Several authors have investigated the formation of composite materials with either pure cellulose or cellulosic materials for the adsorption of heavy metals from water, and some of their findings are presented in this section.

Cellulose-magnetite composites were synthesized for aqueous Cr(VI) adsorption by [53] and [56]. The nanoparticles with sizes ranging between 10 and 40 nm were attached by the bacterial cellulose (BC) nanofibrils forming a composite material with saturation magnetization values of 40 emu g^{-1} [53]. The composite was determined to be superparamagnetic, and the observed results were attributed to the small sizes of the composite particles. Response surface methodology (RSM) was used to better understand the influence of the factors and their interactions on Cr(VI) sequestration. Solution pH and its interaction with the adsorbate concentration were the factors that most significantly influenced the adsorption process. The optimum pH for adsorption was determined to be pH 4 from the influence of the factors on the removal efficiency of chromium. XPS analysis pointed to adsorption followed by Cr(VI) reduction to Cr(III) by a heterogeneous redox process as the adsorption mechanism.

Amino-functionalized magnetite-silica-cellulose (Fe_3O_4@SiO_2@cellulose) nanocomposite was prepared in a multi-step synthesis by [56]. The composite preparation proceeded firstly by magnetite nanoparticle synthesis by co-precipitation followed by deposition of silica onto the Fe_3O_4 nanoparticles, and the Fe_3O_4@SiO_2 particles were suspended in a cellulose solution to form Fe_3O_4@SiO_2@cellulose composite. Amino-functionalization was achieved through grafting of glycidyl methacrylate followed by reaction with ethylenediamine. Cr(VI) adsorption studies indicated that the capacity was highly affected by the solution pH as reported in other studies [42, 53]. The adsorbent showed promising results for Cr(VI) adsorption, and reusability tests confirmed its potential to be applied in up to five cycles while retaining its efficiency. Gupta et al. [17] also reported improved adsorption capacities for Cr(III) adsorption after the formation of composites of multiwalled carbon nanotubes and magnetic iron oxide.

Other carbon-based materials that have been used in the formation of composites with magnetite nanoparticles for adsorption include activated carbon [30, 49], starch [3], wheat straw [58], palm shell [24], and pine cone [42]. From the reports, it was established that the presence of Fe_3O_4 nanoparticles in the composites resulted in ease of magnetic retrieval of the spent adsorbent, while the nanoscale sizes of magnetite generally improved the accessible surface areas in the adsorbents thereby improving their efficiency [17]. The functional groups from organic components of the composites contribute greatly to the sequestration of heavy metal pollutants as previously discussed.

Table 2 summarizes the adsorption capacities for some of the composites discussed in this review.

Table 2 Application of magnetite-organic composites in the adsorption of heavy metal (HM) pollutants

Adsorbent	Surface group	Pollutant	Adsorption capacity (mg g^{-1})	Reference
Polyacrylic acid/silane/Fe_3O_4	– COOH	Cr(III)	54.08	[16]
PPy/Fe_3O_4	– NH	Cr(VI)	169.49	[8]
3-MPA SPION	– SO_3H	Cr(VI)	45	[10]
Fe_3O_4@TSC	– COOR	Cr(III)	549.13	[1]
Fe_3O_4@TSC	– COOR	Co(II)	452.50	[1]
Fe_3O_4 NPs@rGO	– COOH/–OH	Cu(II)	2.20	[27]
Fe_3O_4 NPs@rGO-d-PPG	– COO^-	Cu(II)	13.60	[27]
PS/Fe_3O_4/CS-PEI	– NH_2/–OH	Cu(II)	212.30	[64]
Fe_3O_4/CS	– NH_2/–OH	La(III)	342.46	[18]
Biochar	– COO^-	As(V)	3.68	[33]
Chitosan/biochar (CB)	– NH_2/–OH	As(V)	10.6	[33]
MCB	– NH_2/–OH	As(V)	14.93	[33]
Fe_3O_4@SiO_2@PT	– OH	Au(III)	917.43	[14]
Fe_3O_4@SiO_2@PT	– OH	Pd(II)	196.46	[14]
Fe_3O_4-HBC	– OH/–Si–O	As(V)	3.36	[6]
Fe_3O_4-HBC	– OH/–SiO	As(III)	3.07	[6]
BC-Fe_3O_4	– OH/–COC	Cr(VI)	11.56	[53]
Fe_3O_4@SiO_2@cellulose	– NH_2	Cr(VI)	171.5	[56]

4 Conclusion

Heavy metal contamination of ground water poses challenges in environmental management, and strategies to improve the remediation efficiency are greatly desired. The adsorption process provides an alternative to complex treatment strategies. Adsorption provides ease of operation, selectivity, and wide applicability. The use of different adsorbents provides selectivity for pollutants and increased adsorption efficiency. The review established that the formation of composites of various organic materials with Fe_3O_4 nanoparticles provided high affinities for heavy metal pollutants and increased surface areas and magnetic separability which provided efficient remediation. The presence of interfering ions minimally affected the adsorption process owing to high affinity of Fe_3O_4 for the studied pollutants. Although numerous studies on magnetite and its various composites for the adsorption of pollutants from wastewater, most studies utilize synthetic wastewater and are often conducted in batch mode. Reports on the application of these materials in continuous flows reactors using real wastewater are still limited and would be crucial to the applications of these composite materials in industrial applications.

Acknowledgements The authors acknowledge South Africa National Research Foundation (NRF) and the University of the Western Cape for funding this work.

References

1. Alqadami AA, Naushad M, Abdalla MA, Ahamad T, Abdullah Alothman Z, Alshehri SM (2016) Synthesis and characterization of Fe_3O_4 @TSC nanocomposite: highly efficient removal of toxic metal ions from aqueous medium. RSC Adv 6:22679–22689. https://doi.org/10.1039/C5RA27525C
2. Amara D, Margel S (2011) Solventless thermal decomposition of ferrocene as a new approach for the synthesis of porous superparamagnetic and ferromagnetic composite microspheres of narrow size distribution. J Mater Chem 21:15764–15772. https://doi.org/10.1039/c1jm11842k
3. An B, Liang Q, Zhao D (2011) Removal of arsenic(V) from spent ion exchange brine using a new class of starch-bridged magnetite nanoparticles. Water Res 45:1961–1972. https://doi.org/10.1016/j.watres.2011.01.004
4. Attia TMS, Hu XL, Yin DQ (2014) Synthesised magnetic nanoparticles coated zeolite (MNCZ) for the removal of arsenic (As) from aqueous solution. J Exp Nanosci 9:551–560. https://doi.org/10.1080/17458080.2012.677549
5. Badruddoza AZM, Shawon ZBZ, Tay WJD, Hidajat K, Uddin MS (2013) Fe_3O_4/cyclodextrin polymer nanocomposites for selective heavy metals removal from industrial wastewater. Carbohydr Polym 91:322–332. https://doi.org/10.1016/j.carbpol.2012.08.030
6. Baig SA, Sheng T, Sun C, Xue X, Tan L, Xu X (2014) Arsenic removal from aqueous solutions using Fe_3O_4-HBC composite: effect of calcination on adsorbents performance. PLoS ONE 9:e100704. https://doi.org/10.1371/journal.pone.0100704
7. Bhaumik M, Leswifi TY, Maity A, Srinivasu VV, Onyango MS (2011) Removal of fluoride from aqueous solution by polypyrrole/F_3O_4 magnetic nanocomposite. J Hazard Mater 186:150–159. https://doi.org/10.1016/j.jhazmat.2010.10.098

8. Bhaumik M, Maity A, Srinivasu VV, Onyango MS (2011) Enhanced removal of Cr(VI) from aqueous solution using polypyrrole/Fe_3O_4 magnetic nanocomposite. J Hazard Mater 190:381–390. https://doi.org/10.1016/j.jhazmat.2011.03.062
9. Bhowmick S, Chakraborty S, Mondal P, Van Renterghem W, Van den Berghe S, Roman-Ross G, Chatterjee D, Iglesias M (2014) Montmorillonite-supported nanoscale zero-valent iron for removal of arsenic from aqueous solution: kinetics and mechanism. Chem Eng J. https://doi.org/10.1016/j.cej.2013.12.049
10. Burks T, Avila M, Akhtar F, Göthelid M, Lansåker PCC, Toprak MSS, Muhammed M, Uheida A (2014) Studies on the adsorption of chromium(VI) onto 3-mercaptopropionic acid coated superparamagnetic iron oxide nanoparticles. J Colloid Interface Sci 425:36–43. https://doi.org/10.1016/j.jcis.2014.03.025
11. Darezereshki E, Darban A, Abdollahy M, Jamshidi-Zanjani A (2018) Influence of heavy metals on the adsorption of arsenate by magnetite nanoparticles: Kinetics and thermodynamic. Environ Nanotechnol Monit Manag 10:51–62. https://doi.org/10.1016/j.enmm.2018.04.002
12. Dave PN, Chopda LV (2014) Application of iron oxide nanomaterials for the removal of heavy metals. J Nanotechnol
13. Demirbas A (2008) Heavy metal adsorption onto agro-based waste materials: a review. J Hazard Mater 157:220–229. https://doi.org/10.1016/j.jhazmat.2008.01.024
14. Fan R, Min H, Hong X, Yi Q, Liu W, Zhang Q, Luo Z (2019) Plant tannin immobilized $Fe_3O_4@SiO_2$ microspheres: a novel and green magnetic bio-sorbent with superior adsorption capacities for gold and palladium. J Hazard Mater 364:780–790. https://doi.org/10.1016/j.jhazmat.2018.05.061
15. Gonzalez-Moragas L, Yu SM, Murillo-Cremaes N, Laromaine A, Roig A (2015) Scale-up synthesis of iron oxide nanoparticles by microwave-assisted thermal decomposition. Chem Eng J 281:87–95. https://doi.org/10.1016/j.cej.2015.06.066
16. Guan X, Chang J, Chen Y, Fan H (2015) A magnetically-separable Fe_3O_4 nanoparticle surface grafted with polyacrylic acid for chromium(iii) removal from tannery effluents. RSC Adv 5:50126–50136. https://doi.org/10.1039/C5RA06659J
17. Gupta VK, Agarwal S, Saleh TA (2011) Chromium removal by combining the magnetic properties of iron oxide with adsorption properties of carbon nanotubes. Water Res 45:2207–2212. https://doi.org/10.1016/j.watres.2011.01.012
18. Haldorai Y, Rengaraj A, Ryu T, Shin J, Huh YS, Han Y-K (2015) Response surface methodology for the optimization of lanthanum removal from an aqueous solution using a Fe_3O_4/chitosan nanocomposite. Mater Sci Eng B 195:20–29. https://doi.org/10.1016/j.mseb.2015.01.006
19. Horst MF, Lassalle V, Ferreira ML (2015) Nanosized magnetite in low cost materials for remediation of water polluted with toxic metals, azo- and antraquinonic dyes. Front Environ Sci Eng 9:746–769. https://doi.org/10.1007/s11783-015-0814-x
20. Iconaru SL, Guégan R, Popa CL, Motelica-Heino M, Ciobanu CS, Predoi D (2016) Magnetite (Fe_3O_4) nanoparticles as adsorbents for As and Cu removal. Appl Clay Sci 134:128–135. https://doi.org/10.1016/j.clay.2016.08.019
21. Ihsanullah AA, Al-Amer AM, Laoui T, Al-Marri MJ, Nasser MS, Khraisheh M, Atieh MA (2016) Heavy metal removal from aqueous solution by advanced carbon nanotubes: critical review of adsorption applications. Sep Purif Technol 157:141–161. https://doi.org/10.1016/j.seppur.2015.11.039
22. Inoue K (2011) Heavy metal toxicity. J Clin Toxicol s3:1–2. https://doi.org/10.4172/2161-0495.S3-007
23. Itoh H, Sugimoto T (2003) Systematic control of size, shape, structure, and magnetic properties of uniform magnetite and maghemite particles. J Colloid Interface Sci 265:283–295. https://doi.org/10.1016/S0021-9797(03)00511-3
24. Jais FM, Ibrahim S, Yoon Y, Jang M (2016) Enhanced arsenate removal by lanthanum and nano-magnetite composite incorporated palm shell waste-based activated carbon. Sep Purif Technol 169:93–102. https://doi.org/10.1016/j.seppur.2016.05.034
25. Khalighyaan N, Hooshmand N, Razzaghi-Asl N, Zare K, Miri R (2014) Response surface strategy in the synthesis of Fe_3O_4 nanoparticles. Int J Nano Dimens 5:351–363

26. Khatoon N, Khan AH, Pathak V, Agnihotri N, Rehman M (2013) Removal of hexavalent chromium from synthetic waste water using synthetic Nano Zero Valent Iron (NZVI) as adsorbent. Int J Innov Res Sci Eng Technol 2:6140–6149
27. Kim HJ, Choi H, Sharma AK, Hong WG, Shin K, Song H, Kim HY, Hong YJ (2021) Recyclable aqueous metal adsorbent: synthesis and Cu(II) sorption characteristics of ternary nanocomposites of Fe_3O_4 nanoparticles@graphene–poly-*N*-phenylglycine nanofibers. J Hazard Mater 401:123283. https://doi.org/10.1016/j.jhazmat.2020.123283
28. Kolen'Ko YV, Bañobre-López M, Rodríguez-Abreu C, Carbó-Argibay E, Sailsman A, Piñeiro-Redondo Y, Cerqueira MF, Petrovykh DY, Kovnir K, Lebedev OI, Rivas J (2014) Large-scale synthesis of colloidal Fe_3O_4 nanoparticles exhibiting high heating efficiency in magnetic hyperthermia. J Phys Chem C 118:8691–8701. https://doi.org/10.1021/jp500816u
29. Kumari M, Pittman CU, Mohan D (2015) Heavy metals [chromium (VI) and lead (II)] removal from water using mesoporous magnetite (Fe_3O_4) nanospheres. J Colloid Interface Sci 442:120–132. https://doi.org/10.1016/j.jcis.2014.09.012
30. Kwon JH, Wilson LD, Sammynaiken R (2014) Synthesis and characterization of magnetite and activated carbon binary composites. Synth Met 197:8–17. https://doi.org/10.1016/j.synthmet.2014.08.010
31. Lasheen MR, El-Sherif IY, Sabry DY, El-Wakeel ST, El-Shahat MF (2016) Adsorption of heavy metals from aqueous solution by magnetite nanoparticles and magnetite-kaolinite nanocomposite: equilibrium, isotherm and kinetic study. Desalin Water Treat 57:17421–17429. https://doi.org/10.1080/19443994.2015.1085446
32. Latif A, Sheng D, Sun K, Si Y, Azeem M, Abbas A, Bilal M (2020) Remediation of heavy metals polluted environment using Fe-based nanoparticles: mechanisms, influencing factors, and environmental implications. Environ Pollut 264:114728. https://doi.org/10.1016/j.envpol.2020.114728
33. Liu S, Huang B, Chai L, Liu Y, Zeng G, Wang X, Zeng W, Shang M, Deng J, Zhou Z (2017) Enhancement of As(V) adsorption from aqueous solution by a magnetic chitosan/biochar composite. RSC Adv 7:10891–10900. https://doi.org/10.1039/C6RA27341F
34. Luther S, Borgfeld N, Kim J, Parsons JG (2012) Removal of arsenic from aqueous solution: a study of the effects of pH and interfering ions using iron oxide nanomaterials. Microchem J 101:30–36. https://doi.org/10.1016/j.microc.2011.10.001
35. Mahdavian AR, Mirrahimi MA-S (2010) Efficient separation of heavy metal cations by anchoring polyacrylic acid on superparamagnetic magnetite nanoparticles through surface modification. Chem Eng J 159:264–271. https://doi.org/10.1016/j.cej.2010.02.041
36. Mayo JTT, Yavuz C, Yean S, Cong L, Shipley H, Yu W, Falkner J, Kan A, Tomson M, Colvin VLL, Shiple H, Yu W, Falkner J, Kan A, Tomson M, Colvin VLL (2007) The effect of nanocrystalline magnetite size on arsenic removal. Sci Technol Adv Mater 8:71–75. https://doi.org/10.1016/j.stam.2006.10.005
37. Minitha CR, Suresh R, Maity UK, Haldorai Y, Subramaniam V, Manoravi P, Joseph M, Rajendra Kumar RT (2018) Magnetite nanoparticle decorated reduced graphene oxide composite as an efficient and recoverable adsorbent for the removal of Cesium and strontium ions. Ind Eng Chem Res 57:1225–1232. https://doi.org/10.1021/acs.iecr.7b05340
38. Monárrez-Cordero B, Amézaga-Madrid P, Antúnez-Flores W, Leyva-Porras C, Pizá-Ruiz P, Miki-Yoshida M (2014) Highly efficient removal of arsenic metal ions with high superficial area hollow magnetite nanoparticles synthetized by AACVD method. J Alloys Compd 586:S520–S525. https://doi.org/10.1016/j.jallcom.2012.12.073
39. Ofomaja AE, Naidoo EB, Modise SJ (2009) Removal of copper(II) from aqueous solution by pine and base modified pine cone powder as biosorbent. J Hazard Mater 168:909–917. https://doi.org/10.1016/j.jhazmat.2009.02.106
40. Okoli CP, Ofomaja AE (2019) Development of sustainable magnetic polyurethane polymer nanocomposite for abatement of tetracycline antibiotics aqueous pollution: response surface methodology and adsorption dynamics. J Clean Prod. https://doi.org/10.1016/J.JCLEPRO.2019.01.157

41. Ouma ILA, Mushonga P, Madiehe AM, Meyer M, Dejene FB, Onani MO (2014) Synthesis, optical and morphological characterization of MPA-capped PbSe nanocrystals. Phys B Condens Matter 439:130–132. https://doi.org/10.1016/j.physb.2013.10.057
42. Ouma ILA, Naidoo EB, Ofomaja AE (2017) Iron oxide nanoparticles stabilized by lignocellulosic waste as green adsorbent for Cr(VI) removal from wastewater. Eur Phys J Appl Phys 79:30401. https://doi.org/10.1051/epjap/2017160406
43. Ouma ILA, Naidoo EB, Ofomaja AE (2018) Thermodynamic, kinetic and spectroscopic investigation of arsenite adsorption mechanism on pine cone-magnetite composite. J Environ Chem Eng 6:5409–5419. https://doi.org/10.1016/j.jece.2018.08.035
44. Ouma ILA, Naidoo EB, Ofomaja AE (2019) An insight into the adsorption mechanism of hexavalent chromium onto magnetic pine cone powder. Chemistry for a clean and healthy planet. Springer International Publishing, Cham, pp 185–195
45. Ouma L, Ofomaja A (2020) Probing the interaction effects of metal ions in Mnx Fe(3–x)O_4 on arsenite oxidation and adsorption. RSC Adv 10:2812–2822. https://doi.org/10.1039/C9RA09543H
46. Peng X, Xu F, Zhang W, Wang J, Zeng C, Niu M, Chmielewská E (2014) Magnetic Fe_3O_4 @ silica–xanthan gum composites for aqueous removal and recovery of Pb^{2+}. Colloids Surfaces A Physicochem Eng Asp 443:27–36. https://doi.org/10.1016/j.colsurfa.2013.10.062
47. Pholosi A, Ofomaja AE, Naidoo EB (2013) Effect of chemical extractants on the biosorptive properties of pine cone powder: influence on lead(II) removal mechanism. J Saudi Chem Soc 17:77–86. https://doi.org/10.1016/j.jscs.2011.10.017
48. Ray PZ, Shipley HJ (2015) Inorganic nano-adsorbents for the removal of heavy metals and arsenic: a review. RSC Adv 5:29885–29907. https://doi.org/10.1039/C5RA02714D
49. Salam MA, El-Shishtawy RM, Obaid AY (2014) Synthesis of magnetic multi-walled carbon nanotubes/magnetite/chitin magnetic nanocomposite for the removal of Rose Bengal from real and model solution. J Ind Eng Chem 20:3559–3567. https://doi.org/10.1016/j.jiec.2013.12.049
50. Shafique U, Ijaz A, Salman M, Zaman WU, Jamil N, Rehman R, Javaid A (2012) Removal of arsenic from water using pine leaves. J Taiwan Inst Chem Eng 43:256–263. https://doi.org/10.1016/j.jtice.2011.10.006
51. Shaker S, Zafarian S, Chakra S, Rao VK (2013) Preparation and characterization of magnetite nanoparticles. Int J Innov Res Sci Eng Technol 2:2969–2973
52. Sharma G, Jeevanandam P (2013) Synthesis of self-assembled prismatic iron oxide nanoparticles by a novel thermal decomposition route. RSC Adv 3:189–200. https://doi.org/10.1039/c2ra22004k
53. Stoica-Guzun A, Stroescu M, Jinga SI, Mihalache N, Botez A, Matei C, Berger D, Damian CM, Ionita V (2016) Box-Behnken experimental design for chromium(VI) ions removal by bacterial cellulose-magnetite composites. Int J Biol Macromol 91:1062–1072. https://doi.org/10.1016/j.ijbiomac.2016.06.070
54. Stötzel C, Kurland HD, Grabow J, Müller FA (2015) Gas phase condensation of superparamagnetic iron oxide-silica nanoparticles—control of the intraparticle phase distribution. Nanoscale 7:7734–7744. https://doi.org/10.1039/c5nr00845j
55. Sud D, Mahajan G, Kaur MP (2008) Agricultural waste material as potential adsorbent for sequestering heavy metal ions from aqueous solutions: a review. Bioresour Technol 99:6017–6027. https://doi.org/10.1016/j.biortech.2007.11.064
56. Sun X, Yang L, Li Q, Zhao J, Li X, Wang X, Liu H (2014) Amino-functionalized magnetic cellulose nanocomposite as adsorbent for removal of Cr(VI): synthesis and adsorption studies. Chem Eng J 241:175–183. https://doi.org/10.1016/j.cej.2013.12.051
57. Teja AS, Koh P-Y (2009) Synthesis, properties, and applications of magnetic iron oxide nanoparticles. Prog Cryst Growth Charact Mater 55:22–45. https://doi.org/10.1016/j.pcrysgrow.2008.08.003
58. Tian Y, Wu M, Lin X, Huang P, Huang Y (2011) Synthesis of magnetic wheat straw for arsenic adsorption. J Hazard Mater 193:10–16. https://doi.org/10.1016/j.jhazmat.2011.04.093
59. US EPA (2009) National primary drinking water regulations
60. Volesky B, Holan ZR (1995) Biosorption of heavy metals. Biotechnol Prog 11:235–250

61. Warner CL, Addleman RS, Cinson AD, Droubay TC, Engelhard MH, Nash MA, Yantasee W, Warner MG (2010) High-performance, superparamagnetic, nanoparticle-based heavy metal sorbents for removal of contaminants from natural waters. Chemsuschem 3:749–757. https://doi.org/10.1002/cssc.201000027
62. Warner CL, Chouyyok W, Mackie KE, Neiner D, Saraf LV, Droubay TC, Warner MG, Addleman RS (2012) Manganese doping of magnetic iron oxide nanoparticles: tailoring surface reactivity for a regenerable heavy metal sorbent. Langmuir 28:3931–3937. https://doi.org/10.1021/la2042235
63. Wu L-K, Wu H, Zhang H, Cao H, Hou G, Tang Y, Zheng G-Q (2018) Graphene oxide/$CuFe_2O_4$ foam as an efficient absorbent for arsenic removal from water. Chem Eng J 334:1808–1819. https://doi.org/10.1016/j.cej.2017.11.096
64. Xiao C, Liu X, Mao S, Zhang L, Lu J (2017) Polystyrene/Fe_3O_4/chitosan magnetic composites for the efficient and recyclable adsorption of Cu(II) ions. Appl Surf Sci 394:378–385. https://doi.org/10.1016/j.apsusc.2016.10.116
65. Yan W, Lien H-L, Koel BE, Zhang W (2013) Iron nanoparticles for environmental clean-up: recent developments and future outlook. Environ Sci Process Impacts 15:63. https://doi.org/10.1039/c2em30691c
66. Yeo SY, Choi S, Dien V, Sow-Peh YK, Qi G, Hatton TA, Doyle PS, Thio BJR (2013) Using magnetically responsive tea waste to remove lead in waters under environmentally relevant conditions. PLoS ONE 8:e66648. https://doi.org/10.1371/journal.pone.0066648
67. Zabochnicka-Świątek M, Krzywonos M (2014) Potentials of biosorption and bioaccumulation processes for heavy metal removal. Polish J Environ Stud 23:551–561
68. Zarnegar Z, Safari J (2017) Modified chemical coprecipitation of magnetic magnetite nanoparticles using linear-dendritic copolymers. Green Chem Lett Rev 10:235–240. https://doi.org/10.1080/17518253.2017.1358769
69. Zhang J, Liu W, Zhang M, Zhang X, Niu W, Gao M, Wang X, Du J, Zhang R, Xu Y (2017) Oxygen pressure-tuned epitaxy and magnetic properties of magnetite thin films. J Magn Magn Mater 432:472–476. https://doi.org/10.1016/j.jmmm.2017.02.032
70. Zhou Z, Liu Y, Liu S, Liu H, Zeng G, Tan X (2017) Sorption performance and mechanisms of arsenic (V) removal by magnetic gelatin-modified biochar. Chem Eng J 314:223–231. https://doi.org/10.1016/j.cej.2016.12.113

Zeolite for Treatment of Distillery Wastewater in Fluidized Bed Systems

Seth Apollo and John Kabuba

Abstract Distillery wastewater is among the most polluting industrial effluent because it contains high organic load and significant amount of inorganic pollutants such as phosphates, sulphates, nitrates, chloride, calcium and potassium. Normally biological treatment is applied as the initial treatment step for the degradation of the biodegradable organic compounds. This is complemented by integrating with other treatment methods downstream such as chemical oxidation and physico-chemical processes for further removal of inorganic pollutants and biorecalcitrant organic compounds. Zeolite has favourable adsorption capacity and can be beneficial in improving the performance of integrated biological, chemical and physico-chemical process used in treating distillery wastewater. This work gives an elaborate review on the application of zeolite as a biomass support material in anaerobic digestion and as a catalyst support in photocatalytic treatment of distillery wastewater in fluidized bed reactors. An increase in biomass retention time of 8.5-folds and a low biomass detachment rate of 0.038 d^{-1} has been reported with organic load reduction of above 90% when zeolite is used in anaerobic digesters. Zeolite in anaerobic digester favours the growth of methanogens in the order *Methanococcaceae*, *Methanosarcina* and *Methanosaeta,* respectively. Further to this, factors such as catalyst-to-zeolite ratio and terminal velocity of zeolite particles are significant in the performance of a photocatalytic reactor. This work provides significant information on the use of zeolite to enhance biological and advanced chemical oxidation processes as an integrated approach for the treatment of distillery wastewater.

Keywords Zeolite · Distillery wastewater · Adsorption · Anaerobic digestion · Advanced oxidation processes

S. Apollo · J. Kabuba
Department of Chemical Engineering, Vaal University of Technology, Private Bag X021, Vanderbijlpark, South Africa

S. Apollo (✉)
Department of Physical Sciences, University of Embu, P. O Box 6, Embu 60100, Kenya

E. Lichtfouse et al. (eds.), *Inorganic-Organic Composites for Water and Wastewater Treatment*, Environmental Footprints and Eco-design of Products and Processes,
https://doi.org/10.1007/978-981-16-5916-4_5

1 Introduction

Distillery industries are major environmental pollutants due to their large discharge of wastewater which contain large amounts of organic pollutants as indicated by chemical oxygen demand (COD) value of approximately 80,000 mg/l [1]. Part of the organic pollutants is biorecalcitrant melanoidins which are products of chemical reaction of glucose and amino acids [44]. Anaerobic digestion, due to its biomethane production ability, is considered effective as the first treatment step for distillery effluent [32, 36, 44]. Due to the biorecalcitrant melanodins in the distillery effluent, further treatment methods are used after the anaerobic [39, 44].

Physico-chemical technologies such as adsorption and chemical oxidation methods such as photodegradation have been applied to remove the biorecalcitrant melanoidins from distillery effluent, particularly after the initial biological treatment step [6, 9, 13, 14]. Treatment of distillery effluent by adsorption using cost-effective adsorbents such as natural zeolites has huge potential for colour and melanoidins removal in biologically treated distillery wastewater [37]. Likewise, there is a huge potential for the application of photodegradation because it is a rapid process that do not produce any sludge when used for the degradation of biorecalcitrant organic compounds (Vineetha et al. 2013). The performance of biological, physico-chemical and chemical oxidation processes used to treat distillery effluent can be enhanced by use of naturally occurring adsorbents like zeolite [8, 23, 29].

Naturally occurring zeolites, microporous aluminosilicate minerals, can be used to enhance anaerobic digestion, adsorption and photodegradation processes used in distillery effluent treatment due to their unique adsorption properties. When applied in anaerobic digesters, the microporous zeolites provide conducive sites for microbial growth and also adsorb the organic pollutants hence increasing their contact with microorganisms [18]. When applied in adsorption process, zeolites have huge adsorption capacity for biorecalcitrant melanoidins and inorganic pollutants in distillery wastewater [37]. Finally, zeolites can be used as photocatalyst support material in advanced oxidation process to improve degradation of biorecalcitrant melanoidins through synergistic adsorption photodegradation process [26]. This work reviews the application of natural zeolite in enhancing the efficiency of anaerobic digestion, adsorption and photodegradation applied in the abatement of pollution in distillery wastewater. Further to this, the performance of zeolite in robust fluidized bed reactors applied in adsorption, photodegradation and anaerobic digestion of distillery wastewater is discussed.

Table 1 Review of composition of DWW

Parameter (mg/l)	[43]	[46]	[44]
Colour	Dark brown	–	Dark brown
Temperature (°C)	80–90	–	–
pH	4–4.6	3.75–3.83	3.0–5.4
Conductivity (mS/cm)	26–31	–	–
Total suspended solids	4500–7000	–	350
Chemical oxygen demand	85,000–110,000	68,560–76,600	65,000–130,000
Biochemical oxygen demand	25,000–35,000	29,700–29,800	30,000–70,000
Volatile fatty acids	5,200–8,000	1,500–1,600	–
Sulphate	13,100–13,800	–	2,000–6, 000
Total nitrogen	4,200–4,800	–	1,000–2,000
Chlorides	4,500–8,400	–	–
Phosphates	1,500–2,200	–	800–1,200
Phenols	3,000–4,000	450–460	–

1.1 Distillery Wastewater

1.1.1 Characteristics of Distillery Wastewater

The increase in demand for alcohol due to wide spread application of alcohol as fuel, in food and pharmaceuticals, has led to the rise in distillery industries resulting in a corresponding increase in wastewater discharge [44]. The ratio of alcohol produced to wastewater generated during alcohol manufacturing is reported to be 1:15 [43] The wastewater generated is dark brown in colour and impedes light penetration in receiving water bodies that leads to asphyxia due to impaired photosynthesis [43]. A review of composition of distillery wastewater is shown in Table 1.

Distillery effluent contains biorecalcitrant melanoidins which are responsible for its dark brown colour [16, 34]. It is because the melanoidins pass through biological treatment without being degraded that an integrated approach is necessary to achieve complete treatment [5, 13]. Normally melanoidins are removed through post-treatment methods such as adsorption [37], ozonation [40, 46] and photocatalyitic degradation [12, 41].

1.1.2 Melanoidins in Distillery Effluent

Melanoidins are brown organic compounds that contain nitrogen heteroatom and are produced by Millard reaction between amino acids and simple sugars [49]. Due to their dark colour, they are harmful to the aquatic photosynthetic plants because they hinder light penetration in aquatic environment [19]. The structure of melanoidin produced in the MR heavily depends on the reaction conditions and the type of amino acids and sugars in the reaction [49].

2 Natural Zeolite

Zeolites are mineral material that are deposited in rocks near areas that have experienced volcanic activities; therefore, zeolites occur in many regions globally [24]. Zeolites have attractive physical and chemical properties that give them wide application including in the animals feed, separation processes, drug delivery, pollution control and catalysis [10].

2.1 General Properties and Occurrence of Natural Zeolite

The structure of zeolite consists of cation of alkali or alkali earth metals, aluminate, silicate and water as shown by the following general empirical formula:

$$M_{2/n}.Al_2O_4.xSiO_2.yH_2O \tag{1}$$

where x is 2 or higher and n is valency of cation M. The silicate and aluminate are bonded in the structure of zeolite to form a tetrahydral configuration [23]. Natural zeolites are classified according to the proportion of aluminium and silica in their structures and the types and amount of cations present within their structures. There are various types of zeolite depending on their aluminium and silica content, and they are classified by their aluminium-to-silica ratio [11]. Table 2 shows various kinds of natural zeolites and their typical Si:Al ratios.

Zeolite contains about 10–25 wt% water that when evaporated creates voids for uptake of pollutants [31]. The percentage of water removed can be related to the adsorption capacity of zeolite because pores left by the water are available for adsorption. To improve the adsorptive property of zeolite, its surface can be modified to induce desired surface charges thereby adsorbing the targeted pollutants [37].

Another attractive property of zeolite is its cation-exchange ability. Zeolites are known to have good cation-exchange capacity, a property which depends on its cation form and the proportion of silica and aluminium within its structure. It is the cation that exchanges with other cations in wastewater like ammonium, and the order of cation exchange of zeolite is 2 meq per gram of zeolite [31]. Due to this property, zeolites are effectively applied as adsorbent, catalyst and biomass support in wastewater treatment [4, 26].

Table 2 Classification of zeolites [31]

Zeolite type	Si:Al ratio
Analcite	1.00–3.00
Clinoptilolite	2.92–5.04
Chabazite	1.43–4.18
Edingtonite	1.00–2.00
Erionite	3.05–3.99
Faujasite	1.00–3.00
Ferrierite	3.79–6.14
Heulandite	2.85–4.31
Laumontite	1.95–2.25
Mordenite	4.19–5.79
Natrolite	1.5
Phillipsite	1.45–2.87
Stibilite	2.50–5.00
Wairakite	2

2.2 Zeolite as a Suitable Biomass Carrier Material

Zeolites have been widely used as biomass carrier because of their porous structure that gives them large surface area in the range of 25–27 m^2/g and good adsorption capacity of about 23 mg/g as well as their exchangeable cation that contributes good cation-exchange capacity of 2 meq/g [4, 23, 28, 31]. These qualities make zeolite to be an effective biomass support material in anaerobic digesters. The high surface area ensures high attached biomass density while the good adsorptive property concentrates pollutants in close proximity to the biomass for more effective degradation [7, 17]. Also, inhibitory cations like ammonium and metals can be simultaneously removed by zeolite through ion exchange leading to stability in the digester [47].

It is important to determine the type of bacteria which favourably colonizes zeolite surface for better understanding and control of the digestion process. The bacterial strands are identified using scanning electron microscope (SEM) integrated with single-strand conformation polymorphism (SSCP) analysis based on amplification of bacterial and archaeal 16 s rRNA fragments [50]. Methanogens were found to be the bacterial species that mostly colonize the zeolite surface during anaerobic digestion process [31]. Methanogens are the most delicate bacterial strain in the anaerobic digestion process and are very sensitive to changes in the digester operation conditions, and their ability to colonize zeolite surface could lead to an increase in their growth which results in an increase in biogas production [33]. This is due to the fact that acidogens are more resistant types of bacteria in the digester, and they produce volatile fatty acids which should be adequately converted to methane by the methanogens to prevent acidification of the digester and subsequent failure [33]. Zeolite in the reactor therefore ensures the balance in the population of the acid producing methanogens and the methane producing methanogens. Among the methanogens in anaerobic digestion, it was found that the surface of zeolite was rich in *Methanococcaceae* followed by *Methanosarcina* while the least species was *Methanosaeta* [31]. The increase in biogas production when zeolite is used in anaerobic digestion of various substrates has been reported [31].

2.3 Hydrodynamics of Zeolite in Fluidised Bed Reactor

Solid–liquid fluidization is attained when particles in a fluidized bed reactor are made to be suspended in the fluid by a fluid flowing at a superficial liquid velocity above their minimum fluidization velocity (U_{mf}) and below their settling velocity (U_t) [23]. The U_{mf} is the velocity at which the bed undergoes transition from fixed bed to fluidized bed while at velocity above U_t particle entrainment by the upward flowing liquid is experienced. The superficial velocity depends on reactor design and particle properties. For example, larger particles need higher fluidization velocity which increases operation cost.

The best fluidization conditions of zeolite particles in fluidized bed reactor can be predicted by determining U_{mf} and U_t of different sizes of zeolite particles. The U_{mf} is predicted by a semi-empirical correlation form of modified Ergun equation correlating pressure drop and fluid velocity at incipient fluidization. Terminal velocity is calculated using Richardson and Zaki equation which correlates superficial velocity (U) to terminal velocity (U_t) and bed porosity of expanded bed (ε) as [20]:

$$U = U_t \varepsilon^n \tag{2}$$

The bed porosity (ε) which is the fraction of the bed not occupied by the particles is expressed as [23]:

$$\varepsilon = 1 - \frac{m}{\rho_p A_c H} \tag{3}$$

where m is the amount of particles, A_c is the reactor cross sectional area, ρ_p is particle density, and H is bed height. By plotting linearized form of Eq. (1), different hydrodynamic parameters of various sizes of zeolite were determined (Table 3).

2.4 Performance of Zeolite in Anaerobic Fluidised Bed Reactor

Performance analysis of zeolite in anaerobic fluidized bed reactor treating wastewater has been carried out under various hydrodynamic conditions including bed expansion

Table 3 Hydrodynamic characteristics of zeolite particles

Zeolite size (mm)	Reactor size (L)	Static bed height/reactor height	*n* value	U_{mf} (cm/min)	U_t (cm/min)	References
0.7	0.075	0.33	0.305	13.8	792	[23]
2.2	0.075	0.39	0.309	336	336	[23]
0.25–0.315	0.638	0.35	8.67			[22]
0.315–0.5	0.638	0.35	11.3			[22]
0.93	15.7	0.126	3.7	19.2		[21]

and particle size (D_p) [22]. Fernández et al. [17] found that when zeolite of particle sizes 0.2–0.5 mm and 0.5–0.8 mm was used in fluidized bed reactor, increasing bed expansion from 20 to 40% at constant organic loading rate had negligible effect on chemical oxygen demand reduction while there was significant difference in methane production. This is an indication that bed expansion might have influence on methanogens population. Slightly higher performance was obtained at higher bed expansion due to adequate mixing. In the same study, zeolite particles with larger diameter achieved 5% more COD reduction than those with smaller particle diameter, with nearly similar methane production for both. Because of the slight difference in COD reduction, it may be economical to use zeolite with smaller diameter due to ease of fluidization.

Andalib et al. [4] used zeolite in anaerobic fluidized bed reactor to treat thin stillage with high concentration (130,000 mg TCOD/l). Zeolite used had particles size (D_p) of 0.425–0.61 mm with density of 2360 kg/m^3 and settling velocity of 2.8 cm/s while the reactor was operated at up flow velocity of 1.4 cm/s with zeolite load of 0.18 kg/L and a working volume of 16 L. The mean biomass attachment was 37 mg/L with detachment rate of 0.038 d^{-1} corresponding to solid residence time of 30 days which was 8.5 times higher than the hydraulic retention time. This showed that zeolite is good for biomass retention in the reactor. Due to the high solid residence time attained COD removal of up to 88% was achieved. The amount of biogas produced was 16 L/$L_{reactor}$.d which adequately supplemented fluidization at superficial gas velocity of 0.02 cm/s [4].

In another study treating distillery wastewater, it was found that zeolite of particle sizes between 0.25 and 0.8 mm had attached biomass of 40–45 volatile solids (g/L) [17]. Analysis of the colonies using fluorescence in situ hybridization (FISH) identified methanogens of species *Methanosaeta* and *Methanosarcinaceae.* Methane production and COD reduction of 0.3 L CH_4/g COD and 90%, respectively, were recorded at organic loading rate of 20 gCOD/m^{3}d when bed expansion of 25% was used. These findings were in agreement with that of a research by Montalvo et al. [29] in anaerobic digestion treatment of winery wastewater using zeolite in fluidized bed reactor operating at bed expansion of 25%. Attached biomass concentration was found to be 40–46 volatile solids (g/L) corresponding to a COD reduction efficiency of 86%.

Despite the effectiveness of zeolite in enhancing the anaerobic digestion of distillery wastewater by increasing biogas production rate and COD reduction efficiency, the degradation of biorecalcitrant melanoidins is still not achieved [5]. This is because melanoidins pass through biological treatment without being degraded [44]. It has been reported that integrating anaerobic digestion with advanced oxidation processes such as photocatalytic degradation, ozonolysis, among others, can achieve total elimination of organic compounds in distillery wastewater [7, 38, 51].

3 Photocatalytic Degradation

Photocatalytic degradation uses light and semiconductor catalyst for rapid degradation of biorecalcitrant organic compounds in wastewater [41]. Photocatalytic degradation involves the use of a light irradiated on a semiconductor photocatalyst. The irradiated light should be of adequate energy to enable electrons to move from the valence band to the conduction band to initiate reaction with organic pollutants either through the positive hole created in the valence band or the electron in the conduction band. The positive hole can directly oxidize the organic pollutants to mineralization or can oxidize water to produce hydroxyl radical which participate in the oxidation of the organic pollutants [26].

Titanium dioxide (TiO_2) irradiated with UV light or sunlight has been used as a preferred photocatalyst due to its high photosensitivity, environmental safety and stability [21]. Despite the enormous benefits of TiO_2 photocatalyst, it has a main limitation owed to its low density that makes it difficult to separate from treated wastewater through sedimentation process. Because of this, various researchers have suggested methods of attaching the catalyst on various support materials not only to enhance the post-treatment separation efficiency but also to enhance its catalytic property [3, 45].

3.1 Support Materials for TiO_2 Photocatalyst

Over the past few years there has been an exponential growth in interest in the use of TiO_2 in the degradation of biorecalcitrant organic compounds in wastewater [35, 45]. Despite the good performance recorded, the application of TiO_2 in slurry form phases the challenge of recovery from the treated water. The recovery may require additional treatments due to the poor settleability of TiO_2 particles, and this may increase the wastewater treatment cost [45]. To solve this set back, much effort has been focused on developing a titania supported on an appropriate material that can offer high efficiency combined with better recovery [2]. Generally, an ideal support for photocatalyst must have adequate binding between the catalyst and the material; moreover, the attachment should not lead to deactivation of the catalyst active sites; also, the support should have high specific surface area [25, 42].

3.2 Natural Zeolite as Photocatalyst Support Material

Titanium dioxide has previously been attached on non-porous materials like glass beads [25]. However, attaching TiO_2 on porous adsorbent materials like zeolite has gained fame because this integrates the adsorptive property of the material and the catalytic activity of TiO_2. The preparation method of the catalyst support material should be carefully chosen to supress the interaction of catalyst and the active sites of the adsorbent material to maximize adsorption [15].

3.3 Methods of Immobilizing TiO_2 on Zeolite

Supporting TiO_2 onto zeolite results in an effective composite adsorbent-photocatalytic material particularly if the interaction between the two materials do not result in a reduction in the photosensitization of TiO_2 and blockage of the adsorption sites of the zeolite material [15]. The ratio of TiO_2 to zeolite remains a very significant parameter under consideration when preparing the catalyst. When TiO_2 is available on the zeolite surface at low amounts adsorption dominates over photocatalysis, however, at some sufficiently higher concentration of TiO_2, photocatalysis becomes the dominating process. This highlights the need to establish the optimal proportion of zeolite and TiO_2 in the composite material that attains best synergy between adsorption and photodegradation.

Moreover, the calcination temperature employed when attaching TiO_2 on zeolite is also a very significant parameter and can greatly influence the ultimate performance of the photocatalyst. While zeolite is known for its thermal stability, high temperatures can change the physical properties of TiO_2. High calcination temperature has benefit that it leads to the formation of the more catalytically active TiO_2 anatase phase; however, it may lead to conglomeration of TiO_2 particles thus reducing their surface area [21]. Different authors have suggested different calcination temperatures, for example [15] proposed a calcination temperature of 450 °C for 6 h while [21] proposed temperatures of 200 °C for 4 h. Some of the established methods employed in the TiO_2–zeolite composite catalyst preparation are discussed below.

3.3.1 Sol–gel Method

A synthesis pathway where TiO_2 precursor is made into colloids and dispersed in a medium in which support material has been added. In this method, the viscosity of the colloid can be controlled to achieve either thick or thin coating of the catalyst on the support. The TiO_2 precursors used in this method can be titanium alkoxides or titanium halogenides, and desired temperature is used to hydrolyse the precusor and to ensure adequate attachment to the support material to avoid detachment due to attrition or abrasion during water treatment process [26]. During the heating process, hydrolysis through loss of OH group from the precursor and or support material creates an oxygen bridge which enhances attachment of the TiO_2 on the support material [3]. The merit of this synthesis technique is that it ensures strong catalyst adherence. However, the set back of this method is that there is a wide variation of catalyst distribution on the support. It is reported that the calcination temperature and ration of TiO_2 to zeolite have great influence on the effectiveness of a composite catalyst prepared through sol–gel method [26].

3.3.2 Thermal Treatment (Solid–solid Dispersion Method)

In this method, the TiO_2 is mixed with the support material using an appropriate solvent which is then evaporated. This makes the catalyst to physically attach to the support, and the TiO_2 support is then heated at high temperatures in an inert atmosphere [8, 15]. The thermal treatment is a convenient way to ensure adherence and to manipulate the quality of the material by controlling crystal structure, porosity and specific surface area which results into enhanced catalyst activity. As compared to other methods, this method is advantageous since the TiO_2 is employed directly without using any precursor which will need to be crystallized in the process [45].

3.3.3 Chemical Vapour Decomposition Method

This method is convenient for supporting TiO_2 onto a large surface in a short time regardless of the shape of the surface while ensuring high purity because the vapour of material of interest is deposited or condensed on the material. In this method, a support material is exposed to gaseous volatile TiO_2 precursor at a controlled high temperature and pressure in an inert atmosphere [25]. This results in deposition of a desired thin film of TiO_2 particles onto the support surface. Though efficient, this method is expensive and involves a complex procedure.

3.4 Zeolite/TiO_2 Fluidized Bed Reactor

Zeolite in fluidized bed has been used in adsorption, anaerobic treatment and photocatalysis [17, 21, 23]. In photodegradation, titanium dioxide is coated on zeolite to improve its performance because zeolite has high adsorption capacity towards organic pollutants [15]. It is also used to improve resistance to catalyst attrition during fluidization [48] and to improve the efficiency of catalyst separation from the treated water through sedimentation [26]. In a study that prepared TiO_2 photocatalyst immobilized on zeolite through sonication followed by calcination at 450 °C, it was found that the minimum fluidization velocity of the composite catalyst particles was 0.58 cms^{-1} while terminal velocity was 8.3 cms^{-1} [48]. The study further established that a TiO_2 loading of less than 5 wt% led to a growth of longer anatase crystals while increasing the TiO_2 loading between 2.5 and 5 wt% only increased TiO_2 coverage on zeolite by ~23%.

4 Conclusion

This work reviewed the potential use of zeolite in the treatment of distillery wastewater that in nature contains both biodegradable and biorecalcitrant organic compounds together with various types of inorganic pollutants. The variant types of pollutants in the distillery effluent make it impossible to be adequately be treated using a stand-alone process. Therefore, an integrated approach using anaerobic digestion, adsorption and or photodegradation is a practical approach. The review has shown the potential of zeolite to improve the three processes that can be used to effectively handle distillery effluent. Zeolite has adsorptive sites on its surface which are conducive for the immobilization of biomass during anaerobic digestion which improves the efficiency of the biomethanation process. Also, good adsorption capacity of zeolite makes it a good support material for TiO_2 photocatalyst used in the degradation of biorecalcitrant organic compounds in distillery effluent. The efficiency of the photodegradation process depends on the ratio of the zeolite to the catalyst. Factors such as particle size and bed expansion have been shown to have significant contribution in the performance of zeolite in fluidized bed reactors.

References

1. Acharya BK, Mohana S, Madamwar D (2008) Anaerobic treatment of distillery spent wash: a study on upflow anaerobic fixed film bioreactor. Bioresour Technol 99:4621–4626. https://doi.org/10.1016/j.biortech.2007.06.060
2. Akach J, Ochieng A (2017) Chemical engineering research and design Monte Carlo simulation of the light distribution in an annular slurry bubble column photocatalytic reactor. Chem Eng Res Des 129:248–258. https://doi.org/10.1016/j.cherd.2017.11.021
3. Alinsafi A, Evenou F, Abdulkarim EM, Pons MN, Zahraa O, Benhammou A, Yaacoubi A, Nejmeddine A (2007) Treatment of textile industry wastewater by supported photocatalysis. Dye Pigment 74:439–445. https://doi.org/10.1016/j.dyepig.2006.02.024
4. Andalib M, Hafez H, Elbeshbishy E, Nakhla G, Zhu J (2012) Treatment of thin stillage in a high-rate anaerobic fluidized bed bioreactor (AFBR). Bioresour Technol 121:411–418. https://doi.org/10.1016/j.biortech.2012.07.008
5. Apollo S, Aoyi O (2016) Combined anaerobic digestion and photocatalytic treatment of distillery effluent in fluidized bed reactors focusing on energy conservation. Environ Technol (United Kingdom). https://doi.org/10.1080/09593330.2016.1146342
6. Apollo S, Onyango MS, Ochieng A (2013) An integrated anaerobic digestion and UV photocatalytic treatment of distillery wastewater. J Hazard Mater. https://doi.org/10.1016/j.jhazmat.2013.06.058
7. Apollo S, Onyango MS, Ochieng A (2016) Modelling energy efficiency of an integrated anaerobic digestion and photodegradation of distillery effluent using response surface methodology. Environ Technol 3330:1–12. https://doi.org/10.1080/09593330.2016.1151462

8. Apollo S, Onyango S, Ochieng A (2014) UV/H_2O_2/TiO_2/Zeolite Hybrid System for treatment of molasses wastewater. Iran J Chem Chem enginering 33:107–117
9. Arimi MM, Zhang Y, Götz G, Geißen S-U (2015) Treatment of melanoidin wastewater by anaerobic digestion and coagulation. Environ Technol 36:2410–2418. https://doi.org/10.1080/09593330.2015.1032366
10. Bacakova L, Vandrovcova M, Kopova I, Jirka I (2018) Applications of zeolites in biotechnology and medicine: a review. Biomater Sci. https://doi.org/10.1039/C8BM00028J
11. Baker HM, Ghanem R (2009) Evaluation of treated natural zeolite for the removal of o-chlorophenol from aqueous solution. Desalination 249:1265–1272. https://doi.org/10.1016/j.desal.2009.02.059
12. Benton O, Apollo S, Naidoo B, Ochieng A (2016) Photodegradation of molasses wastewater using TiO_2-ZnO nanohybrid photocatalyst supported on activated carbon. Chem Eng Commun 6445:1443–1454. https://doi.org/10.1080/00986445.2016.1201659
13. Chaudhari PK, Mishra IM, Chand S (2007) Decolourization and removal of chemical oxygen demand (COD) with energy recovery: treatment of biodigester effluent of a molasses-based alcohol distillery using inorganic coagulants. Colloids Surf A Physicochem Eng Asp 296:238–247. https://doi.org/10.1016/j.colsurfa.2006.10.005
14. Dolphen R, Thiravetyan P (2011) Adsorption of melanoidins by chitin nanofibers. Chem Eng J 166:890–895. https://doi.org/10.1016/j.cej.2010.11.063
15. Durgakumari V, Subrahmanyam M, Subba RK, Ratnamala A, Noorjahan M, Tanaka K (2002) An easy and efficient use of TiO_2 supported HZSM-5 and TiO_2+HZSM-5 zeolite combinate in the photodegradation of aqueous phenol and p-chlorophenol. Appl Catal A Gen 234:155–165. https://doi.org/10.1016/S0926-860X(02)00224-7
16. Dwyer J, Kavanagh L, Lant P (2008) The degradation of dissolved organic nitrogen associated with melanoidin using a UV/H_2O_2 AOP. Chemosphere 71:1745–1753. https://doi.org/10.1016/j.chemosphere.2007.11.027
17. Fernández N, Montalvo S, Borja R, Guerrero L, Sánchez E, Cortés I, Colmenarejo MF, Travieso L, Raposo F (2008) Performance evaluation of an anaerobic fluidized bed reactor with natural zeolite as support material when treating high-strength distillery wastewater. Renew Energy 33:2458–2466. https://doi.org/10.1016/j.renene.2008.02.002
18. Fernández N, Montalvo S, Fernández-Polanco F, Guerrero L, Cortés I, Borja R, Sánchez E, Travieso L (2007) Real evidence about zeolite as microorganisms immobilizer in anaerobic fluidized bed reactors. Process Biochem 42:721–728. https://doi.org/10.1016/j.procbio.2006.12.004
19. Fuess LT, Garcia ML (2014) Implications of stillage land disposal: a critical review on the impacts of fertigation. J Environ Manage 145:210–229. https://doi.org/10.1016/j.jenvman.2014.07.003
20. Gallant J, Prakash A, Hogg LEW (2011a) Fluidization and hydraulic behaviour of natural zeolite particles used for removal of contaminants from wastewater. Can J Chem Eng 89:159–165. https://doi.org/10.1002/cjce.20391
21. Huang M, Xu C, Wu Z, Huang Y, Lin J, Wu J (2008) Photocatalytic discolorization of methyl orange solution by Pt modified TiO_2 loaded on natural zeolite. Dye Pigment 77:327–334. https://doi.org/10.1016/j.dyepig.2007.01.026
22. Inglezakis VJ, Stylianou M, Loizidou M (2010) Hydrodynamic studies on zeolite fluidized beds. Int J Chem React Eng. https://doi.org/10.2202/1542-6580.2282
23. Jovanovic M, Grbavcic Z, Rajic N, Obradovic B (2014) Removal of Cu(II) from aqueous solutions by using fluidized zeolite A beads: hydrodynamic and sorption studies. Chem Eng Sci 117:85–92. https://doi.org/10.1016/j.ces.2014.06.017
24. Kotoulas A, Agathou D, Triantaphyllidou IE, Tatoulis TI, Akratos CS, Tekerlekopoulou AG, Vayenas DV (2019) Zeolite as a potential medium for ammonium recovery and second cheese whey treatment. Water (Switzerland). https://doi.org/10.3390/w11010136
25. Li Puma G, Bono A, Krishnaiah D, Collin JG (2008) Preparation of titanium dioxide photocatalyst loaded onto activated carbon support using chemical vapor deposition: a review paper. J Hazard Mater 157:209–219. https://doi.org/10.1016/j.jhazmat.2008.01.040

26. Liu ZF, Liu ZC, Wang Y, Li YB, Qu L, Ya J, Huang PY (2012) Photocatalysis of TiO_2 nanoparticles supported on natural zeolite. Mater Technol 27:267–271. https://doi.org/10.1179/1753555712Y.0000000011
27. Mane JD, Modi S, Nagawade S, Phadnis SP, Bhandari VM (2006) Treatment of spentwash using chemically modified bagasse and colour removal studies. Bioresour Technol 97:1752–1755. https://doi.org/10.1016/j.biortech.2005.10.016
28. Milán Z, Sánchez E, Weiland P, Borja R, Martín A, Ilangovan K (2001) Influence of different natural zeolite concentrations on the anaerobic digestion of piggery waste. Bioresour Technol 80:37–43. https://doi.org/10.1016/S0960-8524(01)00064-5
29. Montalvo S, Borja R, Guerrero L, Sa E, Colmenarejo MF, Travieso L, Raposo F (2008a) Performance evaluation of an anaerobic fluidized bed reactor with natural zeolite as support material when treating high-strength distillery wastewater. 33:2458–2466. https://doi.org/10.1016/j.renene.2008.02.002
30. Montalvo S, Guerrero L, Borja R, Cortés I, Sánchez E, Colmenarejo MF (2008) Treatment of wastewater from red and tropical fruit wine production by zeolite anaerobic fluidized bed reactor. J Environ Sci Health B 43:437–442. https://doi.org/10.1080/10934520701796150
31. Montalvo S, Guerrero L, Borja R, Sánchez E, Milán Z, Cortés I, Angeles de la Rubia M (2012) Application of natural zeolites in anaerobic digestion processes: a review. Appl Clay Sci 58:125–133. https://doi.org/10.1016/j.clay.2012.01.013
32. Moraes BS, Junqueira TL, Pavanello LG, Cavalett O, Mantelatto PE, Bonomi A, Zaiat M (2014) Anaerobic digestion of vinasse from sugarcane biorefineries in Brazil from energy, environmental, and economic perspectives: Profit or expense? Appl Energy 113:825–835. https://doi.org/10.1016/j.apenergy.2013.07.018
33. Moraes BS, Zaiat M, Bonomi A (2015) Anaerobic digestion of vinasse from sugarcane ethanol production in Brazil: challenges and perspectives. Renew Sustain Energy Rev 44:888–903. https://doi.org/10.1016/j.rser.2015.01.023
34. Naik N, Jagadeesh KS, Noolvi MN (2010) Enhanced degradation of melanoidin and caramel in biomethanated distillery spentwash by microorganisms isolated from mangroves. 1:347–351
35. Nawi MA, Sabar S, Sheilatina (2012) Photocatalytic decolourisation of Reactive Red 4 dye by an immobilised TiO_2/chitosan layer by layer system. J Colloid Interface Sci 372:80–87. https://doi.org/10.1016/j.jcis.2012.01.024
36. Oller I, Malato S, Sánchez-Pérez JA (2011) Combination of advanced oxidation processes and biological treatments for wastewater decontamination–a review. Sci Total Environ 409:4141–4166. https://doi.org/10.1016/j.scitotenv.2010.08.061
37. Onyango M, Kittinya J, Hadebe N, Ojijo V, Ochieng A (2011) Sorption of melanoidin onto surfactant modified zeolite. Chem Ind Chem Eng Q 17:385–395. https://doi.org/10.2298/CICEQ110125025O
38. Otieno B, Apollo S (2020) Energy recovery from biomethanation of vinasse and its potential application in ozonation post-treatment for removal of biorecalcitrant organic compounds. J Water Process Eng. https://doi.org/10.1016/j.jwpe.2020.101723
39. Otieno B, Apollo S, Kabuba J, Naidoo B, Ochieng A (2019) Ozonolysis post-treatment of anaerobically digested distillery wastewater effluent. Ozone Sci Eng. https://doi.org/10.1080/01919512.2019.1593818
40. Otieno B, Apollo S, Kabuba J, Naidoo B, Ochieng A (2019b) Ozonolysis pre-treatment of waste activated sludge for solubilization and biodegradability enhancement. J Environ Chem Eng 7:1–8. https://doi.org/10.1016/j.jece.2019.102945
41. Peña M, Coca M, González G, Rioja R, García MT (2003) Chemical oxidation of wastewater from molasses fermentation with ozone. Chemosphere 51:893–900. https://doi.org/10.1016/S0045-6535(03)00159-0
42. Pozzo RL, Baltanás MA, Cassano AE (1997) Supported titanium oxide as photocatalyst in water decontamination: state of the art. Catal Today 39:219–231. https://doi.org/10.1016/S0920-5861(97)00103-X
43. Sankaran K, Premalatha M, Vijayasekaran M, Somasundaram VT (2014) DEPHY project: distillery wastewater treatment through anaerobic digestion and phycoremediation—a green

industrial approach. Renew Sustain Energy Rev 37:634–643. https://doi.org/10.1016/j.rser.2014.05.062
44. Satyawali Y, Balakrishnan M (2008) Wastewater treatment in molasses-based alcohol distilleries for COD and color removal: a review. J Environ Manage 86:481–497. https://doi.org/10.1016/j.jenvman.2006.12.024
45. Shan AY, Ghazi TIM, Rashid SA (2010) Immobilisation of titanium dioxide onto supporting materials in heterogeneous photocatalysis: a review. Appl Catal A Gen 389:1–8. https://doi.org/10.1016/j.apcata.2010.08.053
46. Siles JA, García-García I, Martín A, Martín MA (2011) Integrated ozonation and biomethanization treatments of vinasse derived from ethanol manufacturing. J Hazard Mater 188:247–253. https://doi.org/10.1016/j.jhazmat.2011.01.096
47. Tada C, Yang Y, Hanaoka T, Sonoda A, Ooi K, Sawayama S (2005) Effect of natural zeolite on methane production for anaerobic digestion of ammonium rich organic sludge. Bioresour Technol 96:459–464. https://doi.org/10.1016/j.biortech.2004.05.025
48. Vaisman E, Kabir MF, Kantzas A, Langford CH (2005) A fluidized bed photoreactor exploiting a supported photocatalyst with adsorption pre-concentration capacity. J Appl Electrochem 35:675–681. https://doi.org/10.1007/s10800-005-1389-1
49. Wang H-Y, Qian H, Yao W-R (2011) Melanoidins produced by the Maillard reaction: structure and biological activity. Food Chem 128:573–584. https://doi.org/10.1016/j.foodchem.2011.03.075
50. Weiß S, Zankel A, Lebuhn M, Petrak S, Somitsch W, Guebitz GM (2011) Investigation of mircroorganisms colonising activated zeolites during anaerobic biogas production from grass silage. Bioresour Technol 102:4353–4359. https://doi.org/10.1016/j.biortech.2010.12.076
51. Yasar A, Ahmad N, Chaudhry MN, Rehman MSU, Khan A (2007) Ozone for color and COD removal of raw and anaerobically biotreated combined industrial wastewater. Polish J Environ Stud 16:289–294

Photodegradation of Emerging Pollutants Using Catalysts Supported in Organic and Inorganic Composite Materials

Maurício José Paz, Suélen Serafini, Heveline Enzweiler, Luiz Jardel Visioli, and Alexandre Tadeu Paulino

Abstract **Issues**: Efficient solution treatment methods are necessary for the abatement of pharmaceuticals and pesticides in water, and mitigate environmental impacts. Chemical, physical and biological water treatment processes have been applied for the removal of emerging pollutants from water. However, these methods are not completely efficient due to the formation of secondary pollution, high cost and time of operation. Advanced oxidation processes can overcome these problems on water and wastewater treatment containing emerging pollutants. **Major advances**: We reviewed in this text catalytic photodegradation processes of pharmaceuticals and pesticides in aqueous media using catalysts incorporated in/on polymer-based porous rigid organic solid supports. Advanced oxidation processes are usually conducted using specific catalysts combined to ultraviolet (UV) radiation emission. Many catalysts have been studied in UV radiation-assisted water treatment techniques, including titanium dioxide (TiO_2), zinc oxide (ZnO), tin dioxide (SnO_2), cerium (IV) dioxide (CeO_2) and tungsten trioxide (WO_3), in addition to chalcogenides (CdS, CdSe). The UV radiation emission with wavelength lower than 385 nm generates

M. J. Paz · H. Enzweiler · A. T. Paulino (✉)
Department of Food and Chemical Engineering, Santa Catarina State University, Br 282, Km 574, Pinhalzinho, SC CEP: 89870-000, Brazil
e-mail: alexandre.paulino@udesc.br

M. J. Paz
e-mail: mauriciojp27@yahoo.com.br

H. Enzweiler
e-mail: heveline.enzweiler@udesc.br

S. Serafini · A. T. Paulino
Postgraduate Program in Food Science and Technology, Santa Catarina State University, Br 282, Km 574, Pinhalzinho, SC CEP: 89870-000, Brazil

L. J. Visioli
Fronteira Sul Federal University, Fronteira Sul, Rd SC 484, Km 02, Chapecó, SC CEP: 89815-899, Brazil

A. T. Paulino
Postgraduate Program in Applied Chemistry, Santa Catarina State University, Rua Paulo Malschitzki, 200, Joinville, SC CEP 89219-710, Brazil

E. Lichtfouse et al. (eds.), *Inorganic-Organic Composites for Water and Wastewater Treatment*, Environmental Footprints and Eco-design of Products and Processes,
https://doi.org/10.1007/978-981-16-5916-4_6

electron–hole pairs on catalyst structures, inducing the generation of free radicals capable of photo-degrading adsorbed pollutants. The photocatalysis of organic pollutants can also take place after emission of either visible light or combination of UV/visible light. The degradation efficiencies can vary from 61.0 to 99.2% depending on the employed system. Many catalysts have low photodegradation efficiency due to their small surface area and low pollutant adsorption capacity. This problem can be overcome with the immobilization of the catalyst in solid rigid supports. Polymer-based porous composite materials have been demonstrated to be potential organic rigid solid supports to improve the photocatalytic degradation efficiency of organic pollutants due mainly to the increase of surface area. In this sense, we have shown the incorporation of metal oxides on polymer-based porous composite materials for the photodegradation of pharmaceuticals and pesticides contained in aqueous solutions.

1 Introduction

The environmental pollution caused by organic and inorganic chemical compounds has increased in the last years due to the increase of anthropological activities around the world [60]. This problem is also result of the lack of appropriate environmental remediation technologies for application on industrial-scale [47]. The industrial, urban, agricultural and animal production activities are significant sources of environmental pollution as they generate many chemical residues with ecotoxicity, affecting the ecosystem and animal/human health [70]. Many regions have been contaminated due to the inappropriate disposal of residues containing emerging pollutants such as pesticides and pharmaceuticals [24].

Emerging pollutants are synthetic, semi-synthetic, organic chemical compounds which have not been commonly detected, identified and quantified in the environment and in biological tissues [23]. Pesticides, herbicides, insecticides, fungicides, antiparasitic and antibiotics are emerging pollutants with high pollution degree which must be removed from the environment to protect humans, animals and plants [51]. The most studied chemical pollutants due to their impacts on the ecosystem and animal/human health include organochlorines, organophosphates, neonicotinoids, pyrethroids, phenylpyrazoles, benzimidazoles, triazoles, triazines, macrocyclic lactones, penicillins, tetracyclines and so forth [63]. These compounds are significant emerging pollutants for the environment due to the bioaccumulation capacity, and acute, chronic ecotoxicological and toxicological effects on non-target organisms, including humans [62].

The intensive use of pharmaceutical compounds for prophylaxis of animals, and the inadequate management of industrial and urban wastewaters are sources of entry of pesticides and medicines in the environment [60]. Hence, no-treated industrial and urban wastewaters increase the environmental pollution indices as some agricultural pesticides and pharmacological compounds can be persistent and mobile, especially in soil and water [46]. The water contamination by emerging pollutants generates great concern since it is a vector to transport chemicals into groundwater over long distances [56]. Moreover, it is not yet common to find conventional methodologies involving water and wastewater treatment plants containing pesticides and pharmaceuticals [10].

New water and wastewater remediation methodologies contaminated with emerging pollutants are mandatory to mitigate damages to the environment, animals and human beings. In this sense, various physical, chemical and biological processes have been studied/applied in the residue remediation containing pesticides and medicines [58]. These types of technologies are being frequently classified as containment-immobilization, separation or destruction [47]. Containment-immobilization technologies are identified by the use of adsorbent compounds and physical barriers, whereas separation technologies are applied by using extracting compounds, solvents, surfactants and washing [60]. Chemical destruction technologies include ionization, hydrolysis or oxidation reactions, whereas biological destruction technologies include the use of microorganisms, microalgae, composting, landfarming, biopiles, slurry bioreactors, vermeremediation and phytoremediation [47].

Many environmental pollution remediation technologies have still limitations for the applications in real-world situations, including generation of secondary pollution due to formation of toxic by-products, undesirable physiochemical and/or biological properties in the treated environmental matrix, and high operation cost, energy consumption and operation time [23]. Moreover, negative influences of temperature, pH, moisture content, nutrients and matrix compounds are also known problems during applications of some technologies. One of the biggest challenges for applications of environmental remediation technologies is the scale-up from a laboratory plant to industrial plant [23, 47, 60]. Hence, new water and wastewater treatment technologies containing pesticides and medicines have been studied to overcome the main problems currently found in the developed methods. Moreover, detailed studies are necessary to optimize the laboratory experiments to full-scale plant aiming to treat water and wastewater contaminated with emerging pollutants such as agricultural pesticides and pharmacological medicines.

2 Methodologies for Water and Wastewater Treatment

Many emerging pollutants contained in water and wastewater industrial are difficult to biodegrade, needing efficient physical, chemical or biological treatment techniques for their elimination. The traditional aqueous solution treatment techniques commonly applied to water and wastewater purification have proved to be little effective as they require physical, physicochemical, or chemical operations to achieve adequate pollutant removal efficiencies. The most common water and wastewater treatment processes include coagulation, flocculation, electroflocculation, membrane-filtration, adsorption, and advanced oxidative processes (AOPs) [13, 48].

The most widely used methods for water and wastewater treatment are coagulation and flocculation. Both experimental procedures are based on agglomeration and growth of contaminating particles suspended in aqueous media for posterior physical removal [58]. Coagulation and flocculation are viable strategies for the water and wastewater treatment due to inexpensive implantation and operation, in addition to be easily applied at full-scale plants. However, they exhibit low removal efficiencies of emergent pollutants such as pesticides and pharmaceuticals (less than 10%) due to physiochemical properties of these compounds [2]. On the other hand, coagulation and flocculation are potential strategies in pre-treatment processes, followed by a more specific technology of emerging pollutant abatement in water [46, 71].

Electroflocculation and electrocoagulation as well as conventional flocculation and coagulation are techniques employed to agglutinate contaminants contained in aqueous solutions for posterior physical removal. However, processes using electrical current have advantages due to electrical charge action in possible oxidation–reduction reactions of compounds, destabilization of emulsions, and formation of flakes with faster kinetics [65]. Electroflocculation and electrocoagulation are relatively efficient methods for the emerging pollutant removal from aqueous solutions as they are versatile and simple to operate. These methodologies can also be important as previous steps during the execution of a conventional process with the aim of decreasing the concentrations of pollutants in solutions [49, 68].

Another method that can be applied to remove emerging pollutants from aqueous solutions is the membrane-filtration technology. This method is based on the physical separation between aqueous solution and pollutant according to selective permeability of the employed membrane. The water and wastewater purification process using membrane-filtration is very efficient, mainly at low pollutant concentrations. This technology is important to applications in urban and industrial wastewaters [33]. Aqueous solution purification processes involving semipermeable polymeric membranes have become a reality for the removal of persistent pollutants due to high retention efficiencies of impurities on such membranes [26]. The main disadvantage of this technology is the need of previous treatment processes before using the membrane-filtration in the removal of solid particles and high pollutant concentrations contained in aqueous samples. High solute concentration in aqueous solution obstruct the membrane, making the treatment process unfeasible [26, 33]. Membrane-filtration technology has also been applied as an important tool in the oxidation of emerging pollutants as the membrane can be an efficient solid support to encapsulate catalysts. In this case, it takes place the pollutant physical sorption and chemical modification by applying photocatalysis techniques [38, 57].

Sorption has shown to be an important phenomenon in aqueous solution purification processes containing low pollutant concentrations. This technique involves a physical removal process of pollutants after their chemical or physical interaction in the structure of the solid material [48]. In this sense, new adsorbent solid materials have been produced to increase the applicability of sorption in the removal of several emerging pollutants with different physiochemical properties [13, 20]. The disadvantage of aqueous solution treatment processes involving sorption is related to production of secondary pollution. It takes place as the pollutant molecule trapped in the adsorbent was only transferred from the aqueous solution to the solid material structure. Therefore, the adsorbent containing pollutant must be appropriately treated before disposing, being necessary to think in alternative processes capable of removing the emerging pollutant without generating secondary pollution [20]. In this sense, advanced oxidation processes can overcome this disadvantage.

Advanced oxidation processes (AOPs) are based on oxidation reactions of harmful chemical compounds contained in aqueous solutions with subsequent conversion to no-harmful intermediates. In this case, the pollutant removal from solution takes place by physical process as described in other technologies. The oxidative conversion during an advanced oxidation process can sometimes mineralize pollutant compounds to CO_2 and H_2O or generate partial degradation with formation of oxidized compounds [37]. Briefly, the advanced oxidation process starts with the generation of free radicals, especially hydroxyl and oxygen radicals, which have high oxidation capacity of organic pollutants in solutions [24]. There are several advanced oxidation processes, including electrochemical oxidation, ozonation, fenton process, and photocatalytic degradation.

Electrochemical oxidation processes are conducted with production of free radicals after the application of an electrical potential difference between two electrodes (cathode and anode). In this case, the generated electric current provides necessary energy for the formation of free radicals [43], starting the pollutant degradation process. Ozonation is another strategy for the abatement of pollutants in water. The ozone molecule is naturally a strong oxidant agent, acting directly in pollutant oxidation processes. Moreover, it can be efficiently employed for the free hydroxyl radical ($OH\cdot$) formation [55]. The fenton process is applied for the free radical ($OH\cdot$ e $HO_2\cdot$) formation by using a mixture of iron (II) and iron (III) in solution [21]. The fenton process is homogeneous when iron (II) and iron (III) are dissolved in solution, and heterogeneous when iron (II) e iron (III) are immobilized in solid supports [8]. Finally, photocatalytic degradation is a technique based on the application of electromagnetic radiation as energy source for free radical formation. In this method, the pollutant removal from aqueous solution is performed by using semiconductor materials capable of absorbing electromagnetic radiation, altering their electronic shell and forming free radicals [54, 72]. When the fenton process is combined to electromagnetic radiation, it is named of photo-fenton process [52]. The pollutant photodegradation can be performed by using different energy sources such as electrical energy and electromagnetic radiation [64]. The main advantage of an advanced oxidation process is the possibility of complete degradation of the emerging pollutant in solution compared to the conventional water treatment methods involving physical removal. It avoids secondary pollution as we can see in different technologies. Overall, photocatalytic degradation is a potential strategy for the purification of aqueous solutions contaminated with emergent pollutants.

3 Catalytic Photodegradation

Heterogeneous photodegradation processes are based on the separation of charge carriers in catalyst structures after sorption of electromagnetic radiation. The sorbed energy activates the catalyst, generating an electron–hole pair (e^-/h^+) [32]. After segregation, it takes place migration of charge carriers towards the catalyst surface for the occurrence of oxidation–reduction reactions (Fig. 1).

During this process, oxygen and water free radicals are formed to degrade sorbed pollutants on the catalyst surface. Sometimes, the pollutant is degraded without the need of free radicals [28]. The photodegradation efficiency is affected by the sorbed energy wavelength, separation process of the charge carriers and sorption capacity of pollutants on the catalyst structure [61]. The sorbed energy amount on the catalyst surface per light photon must be equal or superior to energy difference between the material valence band (VB) and conduction band (CB) (band gap of the catalyst, E_{BG}) for obtaining maximum quantum efficiency. Catalysts with band gap higher than 3 eV require ultraviolet radiation to each light photon generates an electron–hole pair [32]. This commonly takes place during photodegradation processes of emergent pollutants using titanium dioxide (TiO_2) as catalyst [6]. The separation process of the charge carriers in the catalyst structure is a critical moment during photocatalysis as the electron–hole pair must remain without changes during the oxidation–reduction reactions [32]. Overall, sorbed pollutant degradation reactions on catalyst surfaces might not occur at high charge carrier recombination speeds, even at appropriate radiation wavelength. Thereby, techniques for the reduction of the catalyst particle sizes and immobilization processes of the catalyst in/on solid rigid supports have been

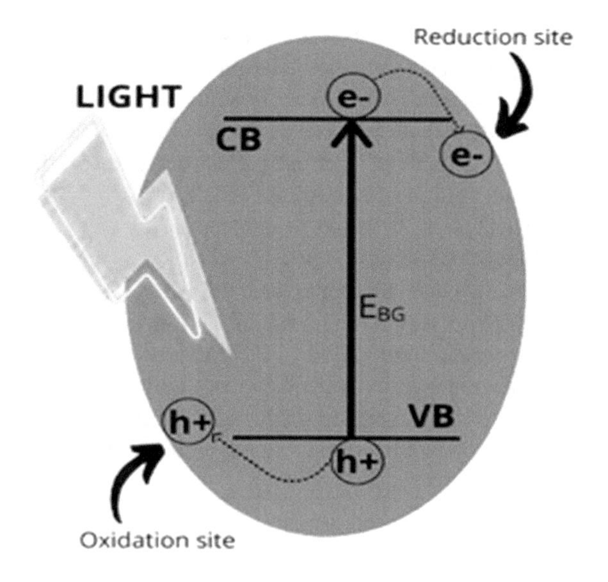

Fig. 1 Schematic drawing of the heterogeneous photodegradation process showing the formation of electron–hole pair (e^-/h^+) and variation of band gap energy (E_{BG}) between the valence band (VB) and conduction band (CB) after electromagnetic radiation emission on the catalyst surface

widely employed to overcome this problem [32]. The compound sorption on catalyst surfaces influences the heterogeneous photodegradation efficiency of emergent pollutants [61]. In this case, the mass transfer process at the interface between catalyst and sorbed pollutant is fundamental for the photodegradation. As the compound sorption on catalyst structures is a slow process compared to recombination speed of charge carriers, the sorption process alters the photodegradation efficiency [61].

The most widely used catalyst in photodegradation processes is the titanium dioxide (TiO_2) as it has low toxicity [29]. The main challenge to use this catalyst is its large band gap energy when the aim is the activation by (solar) visible light. This disadvantage can be overcome by preparing TiO_2 nanostructure with high surface area. This type of nanostructure to allow photodegradation in milder operational conditions, decreasing the cost of the catalytic process [25]. Other metal oxides have also been employed in photocatalysis, including tin dioxide (SnO_2), cerium (IV) oxide (CeO_2) and zinc oxide (ZnO). These compounds are resistant, electromagnetic and low-cost materials when comparing with noble metals [32]. Moreover, ZnO has photocatalytic efficiency similar that found for TiO_2, however, with lower significantly cost [31]. The catalytic activity of ZnO can still be increased after combination with copper oxide II (CuO_2) due to formation of heterojunctions [1]. Metal–organic frameworks (MOFs) are materials formed by transition metals and organic compounds. These materials have proved to be attractive alternatives for the photodegradation processes of emergent pollutants [22]. MOFs and nanocomposites prepared with chalcogenides (CdS, CdSe) are structures capable of absorbing visible radiation due to lower band gap energy [27]. It is very interesting for photocatalytic processes by using solar light.

Heterogeneous photodegradation is an interesting technique for the aqueous solution treatment containing emergent pollutants such as pesticides [40] and pharmaceuticals [72]. Recent works show that the removal of pesticides and medicines from aqueous solutions by using heterogeneous photocatalysis is satisfactory as it is an eco-friendly technology [4]. For instance, heterogeneous photocatalysis can be used for the photocatalytic degradation of methyl parathion by using UV radiation and ZnO-based catalyst [4]. In this case, the morphologic structure modification of the pure zinc oxide from nanospheres to nanorods increased the degradation percentage of methyl parathion from 65 to 98% after 3 h irradiation. Additionally, the operation time was reduced to 80 min by impregnating ZnO in nanorods without efficiency loss. Zirconium-doped TiO_2 thin films proved to be efficient for the chloridazon and 4-chlorophenol photodegradation with radiation of wavelength higher than 320 nm [44]. An almost complete degradation for both compounds can be obtained after 4 h of study. Moreover, several herbicides could be degraded with visible (solar) light radiation by using oxidized carbon nitride as catalyst [36].

Dichlorophenoxyacetic acid (2,4-D) is one of the most widely applied herbicides in agricultural crops. When 2,4-D enter the environment causes toxicity for humans and animals. Its high chemical stability difficult the photodegradation process [34]. Table 1 shows some applications of photocatalytic processes for the degradation of 2,4-D. Several experimental conditions have to be optimized for the efficient abatement of 2,4-D in water, including catalyst amount, initial pollutant concentration,

Table 1 Photocatalytic processes for the degradation of 2,4-dichrolophenoxyacetic acid (2,4-D) with different catalysts and ultraviolet (UV) radiation wavelengths

Photocatalyst	Conditions	Main results	References
Oxygen-doped graphitic carbon nitride	– 15 mg of catalyst (0.06 mg/mL) – Substrate (25 mL, 100 ppm) – UV light irradiation (250 W Hg lamp or the visible light irradiation (a blue light laser)	– Degradation of 98.7% (UV, 140 min) – Degradation of 61.5% (visible, 120 min)	[16]
Fe_3O_4@WO_3/SBA-15	– 40 mg of catalyst – Substrate (100 mL, 10^{-6} mol/L) – UV irradiation (60 W, $\lambda = 254$ nm), 240 min	– Degradation of 90.73%	[34]
Au-TiO_2 [biphasic nanobelt structure (TiO_2(B)/anatase) with anatase as predominant phase]	– Pyrex cylindrical reactor – 1 g/L of catalyst – Substrate (0.53 mmol/L) – pH 3 – Continuous air-bubbling, under UVA radiation (60 W), 120 min	– Degradation of 99.2% (toxicity mainly due intermediate 2,4-dichlorophenol and was eliminated in 4 h)	[7]

wavelength and power of the radiation source, and reaction time. Some catalysts that could be employed for the abatement of 2,4-D in aqueous media by photocatalysis are carbonaceous-based, oxide-based and zeolite-based materials. The catalytic activity increase can be efficiently obtained by doping porous solid supports and changing the catalyst morphologies. Photodegradation experiments of 2,4-D at optimized conditions can achieve a removal efficiency around 90% at short reaction times.

Photocatalytic degradation processes are also useful for aqueous solution treatments containing pharmaceuticals [67], including antibiotics and anti-inflammatories [19]. Caffeine has been utilized as model pollutant in several photodegradation studies due to its excellent stability and toxicity [5, 42], in addition to be present in medicine such as anti-flu. According to [72], less than 40 studies about photocatalytic treatments of residues containing antibiotics were published by 2014, with increase to 280 in 2019. Different strategies have been employed to increase the antibiotic photodegradation efficiencies in solutions aqueous. For example, the levofloxacin and tetracycline removal efficiency from water achieved around 95% using photocatalysis [17, 18]. Anti-inflammatories as the non-steroids are emergent pollutants which are frequently being found in urban aqueous wastewaters due mainly to their random use and disposal. Overall, non-steroids, diclofenac and ibuprofen are the most widely found emerging pollutants in surface water [9, 30]. Table 2 shows a variety of applications for the photocatalytic degradation processes aiming to reduce

Table 2 Photocatalytic degradation processes for the reduction of the concentrations of diclofenac and ibuprofen in aqueous media

Pharmaceutical	Photocatalyst	Conditions	Main results	Reference
Sodium diclofenac	BiOI-Ag_3PO_4 nanocomposite	– Catalyst (0.5 g/L) – Substrate (10 mL, 5 mg/L) – Solar simulator radiation (40 W tungsten lamp) – Time: 60 min	Efficiency of 61%	[45]
Sodium diclofenac	Co_3O_4/WO_3	– Batch reactor – Catalyst (30 mg) – Substrate (50 mL, 15 ppm) – pH 10.7 – Visible light radiation (80 W Hg lamp, cut-off filter) – Time: 180 min	Efficiency of 98.7%	[41]
Sodium diclofenac	Fluorine-doped ZnO	– Batch reactor – Catalyst (1 g/L) – Substrate (10 mg/L) – pH = 6.5 – Solar simulated radiation (1500 W Xe arc lamp) – Accumulated energy (400 kJ/m^2)	Efficiency of 90%	[59]
Ibuprofen	Au/meso-TiO_2	– Simultaneous photocatalytic degradation and water splitting – Catalyst (50 mg) – Substrate (100 mL) – Substrate solution (15 mg/L), – pH from 5 to 9 – Simulated sunlight irradiation (300 W Xe lamp) – Time: 60 min	Complete degradation with production of 94.5 μmol/h/g of H_2	[73]

(continued)

Table 2 (continued)

Pharmaceutical	Photocatalyst	Conditions	Main results	Reference
Ibuprofen	Z-scheme $CdS/Fe_3O_4/TiO_2$ nanocomposite	– Catalyst (0.2 g) – Substrate (200 mL, 10 mg/L) – Visible light radiation (300 W Xe lamp, cut-off filter) – Time: 180 min	Efficiency of 94.2%	[74]

the concentrations of diclofenac and ibuprofen in aqueous media. A similar variability has been found between the pharmaceutical and pesticide photodegradation processes, highlighting the use of solar radiation and the increase of the quantum efficiency of metal oxide-based catalysts.

Even with several studies about photocatalytic degradation for water and wastewater treatment containing emergent pollutants, some difficulties are still found which need to be overcome. Some problems that are being studied to improve the emerging pollutant photocatalytic efficiencies in aqueous solutions include the impossibility of using visible radiation from the solar emission spectrum [30, 50], the aqueous solution matrix effect on the involved photodegradation mechanisms [39] and catalyst types. Studies have shown that the use of catalyst nanoparticles increases the photocatalytic efficiency. However, nanoparticles are difficult to remove from aqueous solutions and increase the cost/benefit of a purification process [12, 31]. In this sense, catalyst incorporation techniques in solid rigid supports are interesting strategies to overcome the separation problems of the catalyst nanoparticles from aqueous solutions. It can enable the scale-up of a photodegradation technology from laboratory plant to full-scale plant.

4 Photocatalysts Supported on Solid Materials

The use of metal oxide nanoparticles as catalysts in photodegradation processes is advantageous as they have higher active surface areas for interaction with pollutants. However, the recovery of these nanoparticles at the end of the photocatalytic process is ineffective [53]. One of the alternatives to facilitate the recovery of photocatalysts is using incorporation processes in solid rigid supports. Photocatalysts incorporated in solid rigid supports are effective due to higher activity in wider pH ranges, higher catalyst stability, higher surface area, and easy of incorporated catalyst separation at the end of the photocatalytic process [11]. Metal oxides have been incorporated in graphene-based, polymer-based, zeolite-based materials to improve the performance of advanced oxidation processes [69]. The potentials of conduction and valence bands of photocatalysts are affected by aqueous solution pH values. These effects decrease when using photocatalysts supported in solid matrices. Hence, active photocatalysts incorporated in porous solid rigid supports have proved to be efficient reagents for photocatalytic processes in the water and wastewater treatment at wider pH ranges [11]. It enables the water and wastewater treatment containing emerging pollutants without adjusting the original solution pH. This is important as some types of emerging pollutants are contained in either acid or alkaline solutions, not being necessary the previous pH adjustment before starting the photodegradation process [6].

The physiochemical properties of the most of solid rigid supports are important for incorporation processes of photocatalysts [6]. For example, the catalytic activity of iron oxide II (FeO) increases when this catalyst is supported on clinoptilolite nanoparticles due to the increase of the active site amounts generating free radicals [3]. The incorporation of either FeO or even other catalyst in semiconductor-based materials has indicated significant improvements of photocatalytic efficiency, variation of the recombination rate between electrons and holes, and changes in the band gap energy [3, 34]. Graphene-based hydrogel nanocomposites are excellent alternatives for the incorporation of palladium due to high reduction capacity of nitroarene in aqueous solutions by photodegradation. In this case, the photodegradation efficiency was higher than 99%, with the advantage of being possible the photocatalyst reutilization in different photocatalytic processes [15]. These types of nanocomposites could be potential alternatives for the incorporation of other catalysts for the photodegradation of emerging pollutants.

Polymeric nanocomposite structures can affect the efficiency of photocatalysts after the incorporation process [27]. Thus, further researches are still needed to comprehend the interaction between solid support and photocatalyst during water and wastewater treatment involving advanced oxidation processes. Studies have highlighted that functional groups such as phosphorus, sulphur and nitrogen present in polymer matrices containing immobilized catalyst affect the heterogeneous catalysis [14]. Overall, catalysts incorporated in either inorganic or organic polymeric composite materials decrease costs of heterogeneous photocatalytic processes as it is possible to recovery the photocatalyst particles for successive studies of emerging pollutant removal contained in aqueous solutions.

Semiconductor oxides doped with known amounts of either transition metals or non-metal species are also potential alternatives for photodegradation studies of emerging pollutants [54]. These types of composites can form electron capture centres, increasing the photocatalytic efficiency and enabling their use for the emerging pollutant removal from aqueous solutions using advanced oxidation processes with application of solar light irradiation. These composites could also be easily incorporated in porous solid rigid supports. However, oxide/polymer structures affect the photocatalytic activity over the supported catalyst. Photocatalysts incorporated in composite membranes have shown excellent results in the aqueous solution purification [35]. Finally, integrated reactors have emerged as an efficient possibility in the purification of water and wastewater containing emerging pollutants. In this system takes place a membrane-filtration process followed by photocatalytic reactions with lower costs for the aqueous solution purification. Filtration membranes containing incorporated photocatalytics are advantageous due to their anti-fouling properties [66].

Acknowledgements ATP thanks the State of Santa Catarina Research and Innovation Foundation (FAPESC, Brazil) for financial support (Grant number: 2019/TR672) and National Council for Scientific and Technological Development (CNPq, Brazil) for the research productivity scholarship (Grant number 312467/2019-2). This study was funded in part by the Coordination for the Advancement of Higher Education Personnel (CAPES, Brazil)—Finance Code 001.

References

1. Acedo-Mendoza AG, Infantes-Molina A, Vargas-Hernández D, Chávez-Sánchez CA, Rodríguez-Castellón E, Tánori-Córdova JC (2020) Photodegradation of methylene blue and methyl orange with CuO supported on ZnO photocatalysts: the effect of copper loading and reaction temperature. Mater Sci Semicond Process 119:105257. https://doi.org/10.1016/j.mssp.2020.105257
2. Ahmed MB, Zhou JL, Ngo HH, Guo W, Thomaidis NS, Xu J (2017) Progress in the biological and chemical treatment technologies for emerging contaminant removal from wastewater: a critical review. J Hazard Mater 323:274–298. https://doi.org/10.1016/j.jhazmat.2016.04.045
3. Arabpour N, Nezamzadeh-Ejhieh A (2015) Modification of clinoptilolite nano-particles with iron oxide: increased composite catalytic activity for photodegradation of cotrimaxazole in aqueous suspension. Mater Sci Semicond Process. https://doi.org/10.1016/j.mssp.2014.12.067
4. Aulakh MK, Kaur S, Pal B, Singh S (2020) Morphological influence of ZnO nanostructures and their Cu loaded composites for effective photodegradation of methyl parathion. Solid State Sci 99:106045. https://doi.org/10.1016/j.solidstatesciences.2019.106045
5. Barrocas B, Chiavassa LD, Conceição Oliveira M, Monteiro OC (2020) Impact of Fe, Mn co-doping in titanate nanowires photocatalytic performance for emergent organic pollutants removal. Chemosphere 250:126240. https://doi.org/10.1016/j.chemosphere.2020.126240
6. Buthiyappan A, Abdul Aziz AR, Wan Daud WMA (2016) Recent advances and prospects of catalytic advanced oxidation process in treating textile effluents. Rev Chem Eng. https://doi.org/10.1515/revce-2015-0034
7. Chenchana A, Nemamcha A, Moumeni H, Doña Rodríguez JM, Araña J, Navío JA, González Díaz O, Pulido Melián E (2019) Photodegradation of 2,4-dichlorophenoxyacetic acid over TiO_2(B)/anatase nanobelts and Au-TiO_2(B)/anatase nanobelts. Appl Surf Sci 467–468:1076–1087. https://doi.org/10.1016/j.apsusc.2018.10.175
8. Cheng M, Zeng G, Huang D, Lai C, Xu P, Zhang C, Liu Y (2016) Hydroxyl radicals based advanced oxidation processes (AOPs) for remediation of soils contaminated with organic compounds: a review. Chem Eng J 284:582–598. https://doi.org/10.1016/j.cej.2015.09.001
9. Chopra S, Kumar D (2020) Ibuprofen as an emerging organic contaminant in environment, distribution and remediation. Heliyon 6:e04087. https://doi.org/10.1016/j.heliyon.2020.e04087
10. Da Le N, Hoang AQ, Hoang TTH, Nguyen TAH, Duong TT, Pham TMH, Nguyen TD, Hoang VC, Phung TXB, Le HT, Tran CS, Dang TH, Vu NT, Nguyen TN, Le TPQ (2020) Antibiotic and antiparasitic residues in surface water of urban rivers in the Red River Delta (Hanoi, Vietnam): concentrations, profiles, source estimation, and risk assessment. Environ Sci Pollut Res. https://doi.org/10.1007/s11356-020-11329-3

11. De Liz MV, De Lima RM, Do Amaral B, Marinho BA, Schneider JT, Nagata N, Peralta-Zamora P (2018) Suspended and immobilized TiO_2 photocatalytic degradation of estrogens: potential for application in wastewater treatment processes. J Braz Chem Soc. https://doi.org/10.21577/0103-5053.20170151
12. Durán A, Monteagudo JM, San Martín I (2018) Operation costs of the solar photo-catalytic degradation of pharmaceuticals in water: a mini-review. Chemosphere 211:482–488. https://doi.org/10.1016/j.chemosphere.2018.07.170
13. Dutt MA, Hanif MA, Nadeem F, Bhatti HN (2020) A review of advances in engineered composite materials popular for wastewater treatment. J Environ Chem Eng 8:104073. https://doi.org/10.1016/j.jece.2020.104073
14. Dzhardimalieva G, Zharmagambetova A, Kudaibergenov S, Igor U (2020) Polymer-immobilized clusters and metal nanoparticles in catalysis. Kinet Catal 61:198–223. https://doi.org/10.1134/S0023158420020044
15. Eghbali P, Nişancl B, Metin Ö (2018) Graphene hydrogel supported palladium nanoparticles as an efficient and reusable heterogeneous catalysts in the transfer hydrogenation of nitroarenes using ammonia borane as a hydrogen source. Pure Appl Chem. https://doi.org/10.1515/pac-2017-0714
16. Ejeta SY, Imae T (2021) Photodegradation of pollutant pesticide by oxidized graphitic carbon nitride catalysts. J Photochem Photobiol A Chem 404:112955. https://doi.org/10.1016/j.jphotochem.2020.112955
17. Fard SG, Haghighi M, Shabani M (2019) Facile one-pot ultrasound-assisted solvothermal fabrication of ball-flowerlike nanostructured (BiOBr)x(Bi7O9I3)1–x solid-solution for high active photodegradation of antibiotic levofloxacin under sun-light. Appl Catal B Environ 248:320–331. https://doi.org/10.1016/j.apcatb.2019.02.021
18. Fattahimoghaddam H, Mahvelati-Shamsabadi T, Lee B-K (2021) Efficient photodegradation of rhodamine B and tetracycline over robust and green g-C_3N_4 nanostructures: supramolecular design. J Hazard Mater 403:123703. https://doi.org/10.1016/j.jhazmat.2020.123703
19. Fauzi AA, Jalil AA, Hitam CNC, Aziz FFA, Chanlek N (2020) Superior sulfate radicals-induced visible-light-driven photodegradation of pharmaceuticals by appropriate Ce loading on fibrous silica ceria. J Environ Chem Eng 8:104484. https://doi.org/10.1016/j.jece.2020.104484
20. Faysal Hossain M, Akther N, Zhou Y (2020) Recent advancements in graphene adsorbents for wastewater treatment: current status and challenges. Chinese Chem Lett. https://doi.org/10.1016/j.cclet.2020.05.011
21. Ganzenko O, Oturan N, Sirés I, Huguenot D, van Hullebusch ED, Esposito G, Oturan MA (2018) Fast and complete removal of the 5-fluorouracil drug from water by electro-Fenton oxidation. Environ Chem Lett 16:281–286. https://doi.org/10.1007/s10311-017-0659-6
22. Gautam S, Agrawal H, Thakur M, Akbari A, Sharda H, Kaur R, Amini M (2020) Metal oxides and metal organic frameworks for the photocatalytic degradation: a review. J Environ Chem Eng 8:103726. https://doi.org/10.1016/j.jece.2020.103726
23. Gavrilescu M, Demnerová K, Aamand J, Agathos S, Fava F (2015) Emerging pollutants in the environment: present and future challenges in biomonitoring, ecological risks and bioremediation. N Biotechnol. https://doi.org/10.1016/j.nbt.2014.01.001
24. Giwa A, Yusuf A, Balogun HA, Sambudi NS, Bilad MR, Adeyemi I, Chakraborty S, Curcio S (2021) Recent advances in advanced oxidation processes for removal of contaminants from water: a comprehensive review. Process Saf Environ Prot 146:220–256. https://doi.org/10.1016/j.psep.2020.08.015
25. Gopinath KP, Madhav NV, Krishnan A, Malolan R, Rangarajan G (2020) Present applications of titanium dioxide for the photocatalytic removal of pollutants from water: a review. J Environ Manage 270:110906. https://doi.org/10.1016/j.jenvman.2020.110906
26. Gurung K, Ncibi MC, Sillanpää M (2019) Removal and fate of emerging organic micropollutants (EOMs) in municipal wastewater by a pilot-scale membrane bioreactor (MBR) treatment under varying solid retention times. Sci Total Environ 667:671–680. https://doi.org/10.1016/j.scitotenv.2019.02.308

27. Hua J, Wang M, Jiao Y, Li H, Yang Y (2018) Strongly coupled CdX (X S, Se and Te) quantum dots/TiO_2 nanocomposites for photocatalytic degradation of benzene under visible light irradiation. Optik (Stuttg). 171:95–106. https://doi.org/10.1016/j.ijleo.2018.06.049
28. Kanakaraju D, Glass BD, Oelgemöller M (2018) Advanced oxidation process-mediated removal of pharmaceuticals from water: a review. J Environ Manage. https://doi.org/10.1016/j.jenvman.2018.04.103
29. Kanakaraju D, Glass BD, Oelgemöller M (2014) Titanium dioxide photocatalysis for pharmaceutical wastewater treatment. Environ Chem Lett 12:27–47. https://doi.org/10.1007/s10311-013-0428-0
30. Kaur A, Umar A, Kansal SK (2016) Heterogeneous photocatalytic studies of analgesic and non-steroidal anti-inflammatory drugs. Appl Catal A Gen 510:134–155. https://doi.org/10.1016/j.apcata.2015.11.008
31. Khan SH, Pathak B (2020) Zinc oxide based photocatalytic degradation of persistent pesticides: a comprehensive review. Environ Nanotechnology Monit Manag 13:100290. https://doi.org/10.1016/j.enmm.2020.100290
32. Kirankumar VS, Sumathi S (2020) A review on photodegradation of organic pollutants using spinel oxide. Mater Today Chem 18:100355. https://doi.org/10.1016/j.mtchem.2020.100355
33. Li C, Cabassud C, Guigui C (2015) Evaluation of membrane bioreactor on removal of pharmaceutical micropollutants: a review. Desalin Water Treat 55:845–858. https://doi.org/10.1080/19443994.2014.926839
34. Lima MS, Cruz-Filho JF, Noleto LFG, Silva LJ, Costa TMS, Luz GE (2020) Synthesis, characterization and catalytic activity of Fe_3O_4@WO_3/SBA-15 on photodegradation of the acid dichlorophenoxyacetic (2,4-D) under UV irradiation. J Environ Chem Eng 8:104145. https://doi.org/10.1016/j.jece.2020.104145
35. Liu L, Liu Z, Bai H, Sun DD (2012) Concurrent filtration and solar photocatalytic disinfection/degradation using high-performance Ag/TiO_2 nanofiber membrane. Water Res. https://doi.org/10.1016/j.watres.2011.12.009
36. Liu X, Li C, Zhang Y, Yu J, Yuan M, Ma Y (2017) Simultaneous photodegradation of multi-herbicides by oxidized carbon nitride: performance and practical application. Appl Catal B Environ 219:194–199. https://doi.org/10.1016/j.apcatb.2017.07.007
37. Luo H, Zeng Y, He D, Pan X (2020) Application of iron-based materials in heterogeneous advanced oxidation processes for wastewater treatment: a review. Chem Eng J. https://doi.org/10.1016/j.cej.2020.127191
38. Luo J, Chen W, Song H, Liu J (2020) Fabrication of hierarchical layer-by-layer membrane as the photocatalytic degradation of foulants and effective mitigation of membrane fouling for wastewater treatment. Sci Total Environ 699:134398. https://doi.org/10.1016/j.scitotenv.2019.134398
39. Mahmoud WMM, Rastogi T, Kümmerer K (2017) Application of titanium dioxide nanoparticles as a photocatalyst for the removal of micropollutants such as pharmaceuticals from water. Curr Opin Green Sustain Chem 6:1–10. https://doi.org/10.1016/j.cogsc.2017.04.001
40. Malakootian M, Shahesmaeili A, Faraji M, Amiri H, Silva Martinez S (2020) Advanced oxidation processes for the removal of organophosphorus pesticides in aqueous matrices: a systematic review and meta-analysis. Process Saf Environ Prot 134:292–307. https://doi.org/10.1016/j.psep.2019.12.004
41. Malefane ME, Feleni U, Kuvarega AT (2020) Cobalt (II/III) oxide and tungsten (VI) oxide p-n heterojunction photocatalyst for photodegradation of diclofenac sodium under visible light. J Environ Chem Eng. https://doi.org/10.1016/j.jece.2019.103560
42. Martín-Gómez AN, Navío JA, Jaramillo-Páez C, Sánchez-Cid P, Hidalgo MC (2020) Hybrid ZnO/Ag3PO_4 photocatalysts, with low and high phosphate molar percentages. J Photochem Photobiol A Chem 388:112196. https://doi.org/10.1016/j.jphotochem.2019.112196
43. Martínez-Huitle CA, Panizza M (2018) Electrochemical oxidation of organic pollutants for wastewater treatment. Curr Opin Electrochem 11:62–71. https://doi.org/10.1016/j.coelec.2018.07.010

44. Mbiri A, Taffa DH, Gatebe E, Wark M (2019) Zirconium doped mesoporous TiO_2 multilayer thin films: influence of the zirconium content on the photodegradation of organic pollutants. Catal Today 328:71–78. https://doi.org/10.1016/j.cattod.2019.01.043
45. Mehrali-Afjani M, Nezamzadeh-Ejhieh A, Aghaei H (2020) A brief study on the kinetic aspect of the photodegradation and mineralization of BiOI-Ag3PO_4 towards sodium diclofenac. Chem Phys Lett. https://doi.org/10.1016/j.cplett.2020.137873
46. Mir-Tutusaus JA, Parladé E, Llorca M, Villagrasa M, Barceló D, Rodriguez-Mozaz S, Martinez-Alonso M, Gaju N, Caminal G, Sarrà M (2017) Pharmaceuticals removal and microbial community assessment in a continuous fungal treatment of non-sterile real hospital wastewater after a coagulation-flocculation pretreatment. Water Res 116:65–75. https://doi.org/10.1016/j.watres.2017.03.005
47. Morillo E, Villaverde J (2017) Advanced technologies for the remediation of pesticide-contaminated soils. Sci Total Environ. https://doi.org/10.1016/j.scitotenv.2017.02.020
48. Najafi H, Farajfaed S, Zolgharnian S, Mosavi Mirak SH, Asasian-Kolur N, Sharifian S (2021) A comprehensive study on modified-pillared clays as an adsorbent in wastewater treatment processes. Process Saf Environ Prot 147:8–36. https://doi.org/10.1016/j.psep.2020.09.028
49. Nidheesh PV, Scaria J, Babu DS, Kumar MS (2021) An overview on combined electrocoagulation-degradation processes for the effective treatment of water and wastewater. Chemosphere 263:127907. https://doi.org/10.1016/j.chemosphere.2020.127907
50. Parul, Kaur K, Badru R, Singh PP, Kaushal S (2020) Photodegradation of organic pollutants using heterojunctions: a review. J Environ Chem Eng 8:103666. https://doi.org/10.1016/j.jece.2020.103666
51. Picó Y, Alvarez-Ruiz R, Alfarhan AH, El-Sheikh MA, Alshahrani HO, Barceló D (2020) Pharmaceuticals, pesticides, personal care products and microplastics contamination assessment of Al-Hassa irrigation network (Saudi Arabia) and its shallow lakes. Sci Total Environ. https://doi.org/10.1016/j.scitotenv.2019.135021
52. Polo-López MI, Sánchez Pérez JA (2020) Perspectives of the solar photo-Fenton process against the spreading of pathogens, antibiotic resistant bacteria and genes in the environment. Curr Opin Green Sustain Chem. https://doi.org/10.1016/j.cogsc.2020.100416
53. Rawtani D, Khatri N, Tyagi S, Pandey G (2018) Nanotechnology-based recent approaches for sensing and remediation of pesticides. J Environ Manage 206:749–762. https://doi.org/10.1016/j.jenvman.2017.11.037
54. Reddy PAK, Reddy PVL, Kwon E, Kim K-H, Akter T, Kalagara S (2016) Recent advances in photocatalytic treatment of pollutants in aqueous media. Environ Int 91:94–103. https://doi.org/10.1016/j.envint.2016.02.012
55. Rekhate CV, Srivastava JK (2020) Recent advances in ozone-based advanced oxidation processes for treatment of wastewater—a review. Chem Eng J Adv 3:100031. https://doi.org/10.1016/j.ceja.2020.100031
56. Rodriguez-Mozaz S, Vaz-Moreira I, Varela Della Giustina S, Llorca M, Barceló D, Schubert S, Berendonk TU, Michael-Kordatou I, Fatta-Kassinos D, Martinez JL, Elpers C, Henriques I, Jaeger T, Schwartz T, Paulshus E, O'Sullivan K, Pärnänen KMM, Virta M, Do TT, Walsh F, Manaia CM (2020) Antibiotic residues in final effluents of European wastewater treatment plants and their impact on the aquatic environment. Environ Int. https://doi.org/10.1016/j.envint.2020.105733
57. Rostam AB, Taghizadeh M (2020) Advanced oxidation processes integrated by membrane reactors and bioreactors for various wastewater treatments: a critical review. J Environ Chem Eng 8:104566. https://doi.org/10.1016/j.jece.2020.104566
58. Rout PR, Zhang TC, Bhunia P, Surampalli RY (2021) Treatment technologies for emerging contaminants in wastewater treatment plants: a review. Sci Total Environ 753:141990. https://doi.org/10.1016/j.scitotenv.2020.141990
59. Rueda-Salaya L, Hernández-Ramírez A, Hinojosa-Reyes L, Guzmán-Mar JL, Villanueva-Rodríguez M, Sánchez-Cervantes E (2020) Solar photocatalytic degradation of diclofenac aqueous solution using fluorine doped zinc oxide as catalyst. J Photochem Photobiol A Chem. https://doi.org/10.1016/j.jphotochem.2020.112364

60. Saleh IA, Zouari N, Al-Ghouti MA (2020) Removal of pesticides from water and wastewater: chemical, physical and biological treatment approaches. Environ Technol Innov. https://doi.org/10.1016/j.eti.2020.101026
61. Salgado BCB, Cardeal RA, Valentini A (2019) Photocatalysis and photodegradation of pollutants. In: Nanomaterials applications for environmental matrices. Elsevier, pp 449–488. https://doi.org/10.1016/B978-0-12-814829-7.00015-X
62. Serafini S, de Freitas Souza C, Baldissera MD, Baldisserotto B, Segat JC, Baretta D, Zanella R, Schafer da Silva A (2019) Fish exposed to water contaminated with eprinomectin show inhibition of the activities of AChE and Na^+/K^+-ATPase in the brain, and changes in natural behavior. Chemosphere. https://doi.org/10.1016/j.chemosphere.2019.02.026
63. Serafini S, Soares JG, Perosa CF, Picoli F, Segat JC, Da Silva AS, Baretta D (2019) Eprinomectin antiparasitic affects survival, reproduction and behavior of Folsomia candida biomarker, and its toxicity depends on the type of soil. Environ Toxicol Pharmacol. https://doi.org/10.1016/j.etap.2019.103262
64. Serpone N, Artemev YM, Ryabchuk VK, Emeline AV, Horikoshi S (2017) Light-driven advanced oxidation processes in the disposal of emerging pharmaceutical contaminants in aqueous media: a brief review. Curr Opin Green Sustain Chem 6:18–33. https://doi.org/10.1016/j.cogsc.2017.05.003
65. Sher F, Hanif K, Iqbal SZ, Imran M (2020) Implications of advanced wastewater treatment: electrocoagulation and electroflocculation of effluent discharged from a wastewater treatment plant. J Water Process Eng 33:101101. https://doi.org/10.1016/j.jwpe.2019.101101
66. Song H, Shao J, Wang J, Zhong X (2014) The removal of natural organic matter with LiCl-TiO_2-doped PVDF membranes by integration of ultrafiltration with photocatalysis. Desalination. https://doi.org/10.1016/j.desal.2014.04.012
67. Tayebee R, Esmaeili E, Maleki B, Khoshniat A, Chahkandi M, Mollania N (2020) Photodegradation of methylene blue and some emerging pharmaceutical micropollutants with an aqueous suspension of WZnO-NH_2@H_3PW12O40 nanocomposite. J Mol Liq 317:113928. https://doi.org/10.1016/j.molliq.2020.113928
68. Tegladza ID, Xu Q, Xu K, Lv G, Lu J (2021) Electrocoagulation processes: a general review about role of electro-generated flocs in pollutant removal. Process Saf Environ Prot 146:169–189. https://doi.org/10.1016/j.psep.2020.08.048
69. Tsang CHA, Li K, Zeng Y, Zhao W, Zhang T, Zhan Y, Xie R, Leung DYC, Huang H (2019) Titanium oxide based photocatalytic materials development and their role of in the air pollutants degradation: overview and forecast. Environ Int. https://doi.org/10.1016/j.envint.2019.01.015
70. Wan Y, Tran TM, Nguyen VT, Wang A, Wang J, Kannan K (2021) Neonicotinoids, fipronil, chlorpyrifos, carbendazim, chlorotriazines, chlorophenoxy herbicides, bentazon, and selected pesticide transformation products in surface water and drinking water from northern Vietnam. Sci Total Environ. https://doi.org/10.1016/j.scitotenv.2020.141507
71. Wang J, Tang X, Xu Y, Cheng X, Li G, Liang H (2020) Hybrid UF/NF process treating secondary effluent of wastewater treatment plants for potable water reuse: adsorption vs. coagulation for removal improvements and membrane fouling alleviation. Environ Res 188:109833. https://doi.org/10.1016/j.envres.2020.109833
72. Yang X, Chen Z, Zhao W, Liu C, Qian X, Zhang M, Wei G, Khan E, Hau Ng Y, Sik Ok Y (2021) Recent advances in photodegradation of antibiotic residues in water. Chem Eng J 405:126806. https://doi.org/10.1016/j.cej.2020.126806
73. Yao X, Hu X, Liu Y, Wang X, Hong X, Chen X, Pillai SC, Dionysiou DD, Wang D (2020) Simultaneous photocatalytic degradation of ibuprofen and H_2 evolution over Au/sheaf-like TiO_2 mesocrystals. Chemosphere. https://doi.org/10.1016/j.chemosphere.2020.127759
74. Zhou A, Liao L, Wu X, Yang K, Li C, Chen W, Xie P (2020) Fabrication of a Z-scheme nanocomposite photocatalyst for enhanced photocatalytic degradation of ibuprofen under visible light irradiation. Sep Pur Technol. https://doi.org/10.1016/j.seppur.2020.117241

Agricultural Wastes Utilization in Water Purification

Anupam Agarwal, Mayuri Rastogi, and N. B. Singh

Abstract Water is considered to be life, and therefore, availability of pure water is a must. Due to industrialization and urbanization, lot of pollutants are discharged into water, which makes water hazardous and harmful to health and environment. Most important pollutants are dyes, organic compounds, and heavy metal ions. Number of methods are being used to purify the water. Out of different methods, adsorption is found to be the most efficient and economical technique. For this method, suitable and low cost adsorbents are required. A pragmatic shift is ongoing in waste material management and wastewater treatment technology due to the large amount of waste production worldwide and the necessity for cheap adsorbents to reduce wastewater treatment costs. A number of agro-industrial wastes and chemically modified wastes are being used as adsorbents for the removal of organic pollutants (dyes and organic compounds) and inorganic pollutants (heavy metals and different ions) from wastewater. Removal efficiencies by chemically modified agro-industrial wastes are found much higher as compared to raw wastes because they have higher surface area and porosity. In this chapter, different agricultural wastes and derived products from agricultural wastes (organic compounds, inorganic compounds, composites, or nanomaterials) have been discussed for decontamination of different pollutants from wastewater. Effect of different parameters on removal efficiency has been described. Adsorption isotherm models and kinetic models have been discussed. Process of desorption and reuse is also pointed out.

1 Introduction

Water and oxygen are most important for survival on Earth. Only 1% of the total water available on earth is accessible as freshwater. Urbanization, industrialization,

A. Agarwal · N. B. Singh (✉)
Department of Chemistry and Biochemistry, SBSR, Sharda University, Greater Noida, India
e-mail: n.b.singh@sharda.ac.in

M. Rastogi
SAHS, Sharda University, Greater Noida, India

E. Lichtfouse et al. (eds.), *Inorganic-Organic Composites for Water and Wastewater Treatment*, Environmental Footprints and Eco-design of Products and Processes,
https://doi.org/10.1007/978-981-16-5916-4_7

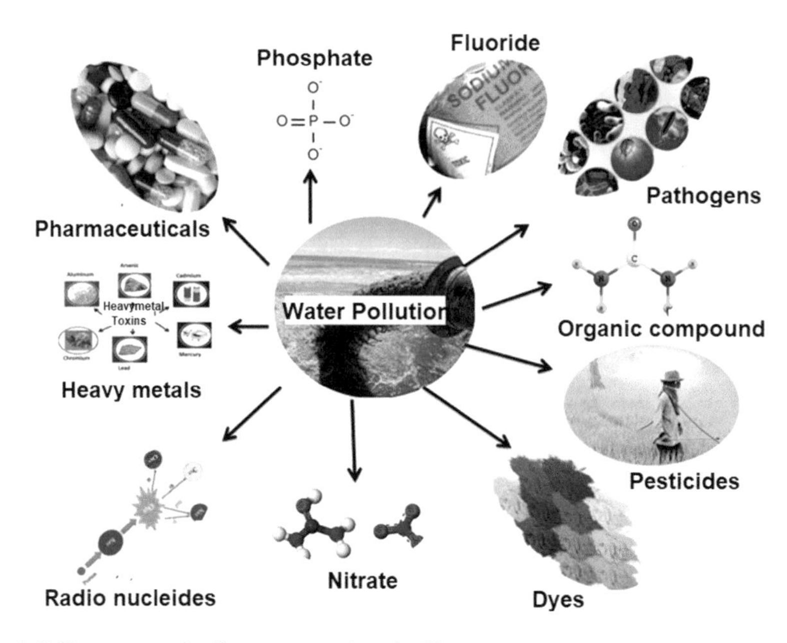

Fig. 1 Different type of pollutants contaminated with water

civilization, and population growth pollute natural resources and become a threat to future generations. Because of this, there is a global scarcity of clean water. There are number of pollutants such as dyes, toxic metals, and microorganisms contaminated with water are very harmful to ecosystem and human health (Fig. 1) [15, 30]. Number of methods used for purification of water is given in Fig. 2.

Out of different methods given in Fig. 2, adsorption technique is found to be the most effective and economical. The efficiency of all adsorbents and wastewater treatment technologies depends on factors like cost-effectiveness, good competence, simplicity of use, equally effective for all kind of pollutants, entails minimum maintenance, fast reaction time and regeneration. Out of different adsorbents, agricultural wastes are most cost effective but the adsorption efficiency may not be very good. In this chapter, adsorbents such as agricultural wastes and products derived from them have been used for purification water and results discussed.

2 Removal of Pollutants by Agriculture Waste

2.1 Raw Agro-Waste and Water Purification

It is reported that orange peel is used for the adsorption of manganese, iron, and copper metal ions from aqueous solution. Surfactant modified orange peel showed

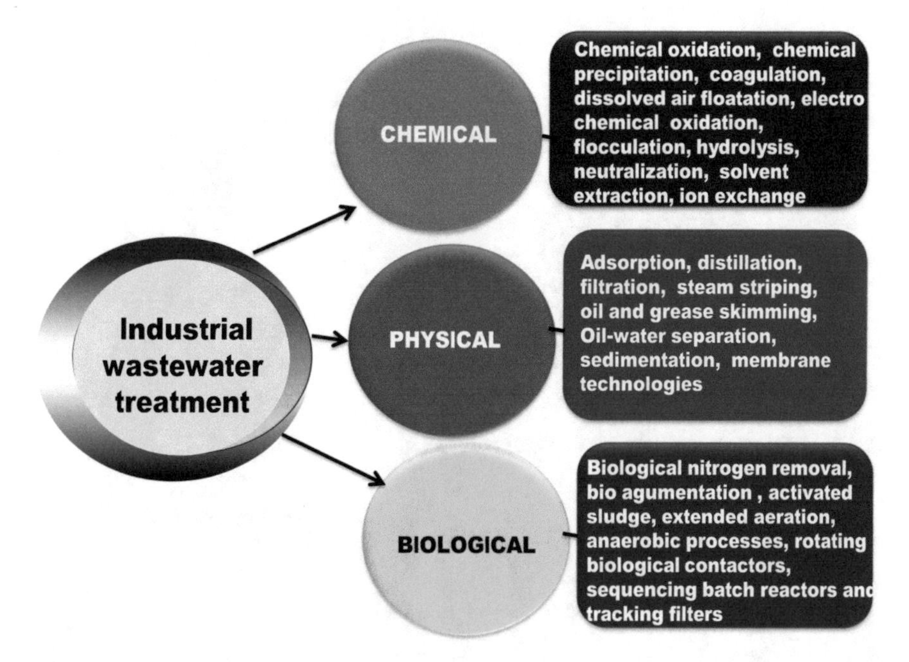

Fig. 2 Different water purification methods

better removal of metal ions. Adsorption is also affected by pH, and the highest adsorption occurred at pH 6.0 [9]. Bamboo shoot skin has been used for the removal of methylene blue dye from aqueous solution, and the maximum adsorption capacity was found to be 29.88 mg/g. Agricultural waste generated from cauliflower leaves was used to prepare bioadsorbent for the removal of methylene blue dye. The leaves were washed, dried, and crushed into powder and used in batch adsorption experiment at different temperatures, concentration of methylene blue dye, adsorbent dose, and pH of the solution. Cauliflower leaves powder can be regenerated up to six cycles by acidic solvent. Maximum adsorption occurs at pH 9.0 [6].

Durian peel, natural product bagasse, coconut shell, walnut shell, castor seeds, palm fruit biomass, rice husk, and sugarcane bagasse are reported as good bioadsorbent for removal of dyes, metal ions, and oil by-products from polluted water (Fig. 3) [8, 16, 20].

Rice husk is considered as a lignocellulosic waste material of low value. It has been reported as a potential adsorbent for removal of cadmium, chromium, and phenolic compounds from water [13]. Rice husk has been used as fixed bed column for the removal of Tl^{3+} from aqueous solution. As the bed height (adsorbent mass) increases, adsorption increases. Breakthrough curves are given for different bed height while other parameters (flow rate 0.4×10^{-3} m^3/min, concentration of thallium nitrate $Tl(NO_3)_3 \cdot 3H_2O = 10$ g/m^3, pH = 10 and T = 298 K) were kept constant (Fig. 4). The results are good for 7 cm height [17].

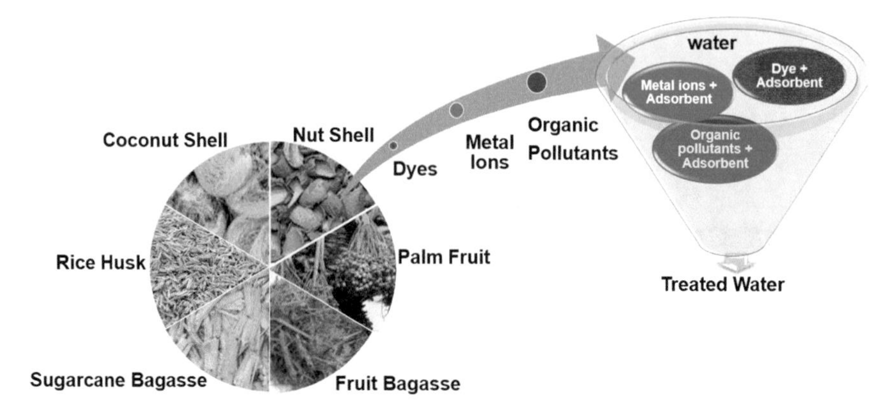

Fig. 3 Different raw agro-wastes for water purification

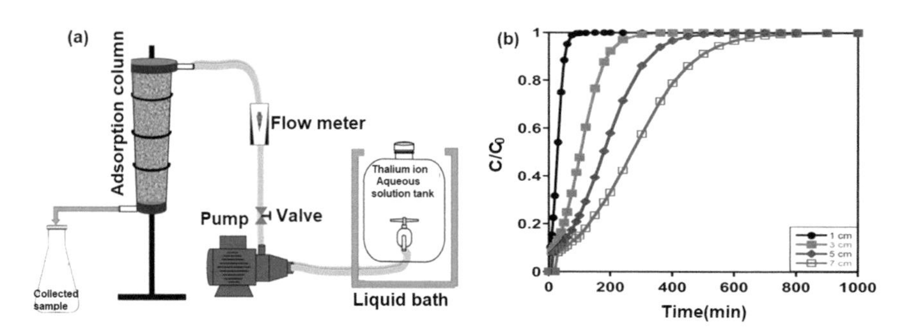

Fig. 4 Adsorption of thallium ion onto rice husk: **a** fixed bed column setup and **b** breakthrough curve [17]

Cucumber peal powder is used as adsorbent for removal of methylene blue dye (Fig. 5) [28]. The adsorbent can be regenerated. The optimum dose of 4 g in 1000 mL

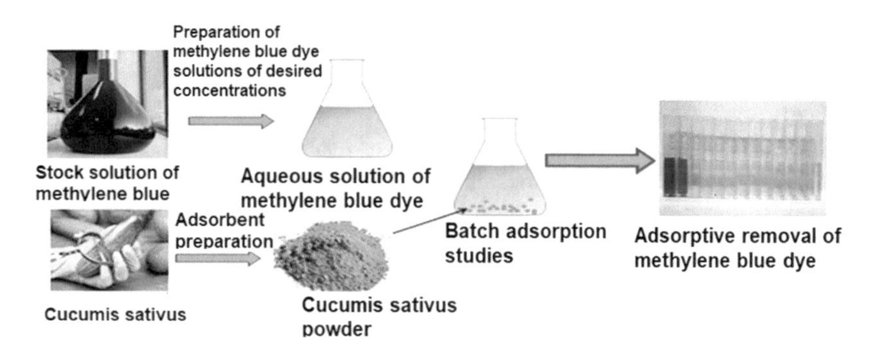

Fig. 5 Removal of methylene blue dye by cucumber peel adsorbent [28]

dye solution (25 mg/L - 250 mg/L concentration) can remove approximately 80% of methylene blue dye in 30 min [28].

Watermelon rind is used for removal of heavy metal ions and dyes from water. The maximum dye adsorption capacity 231.48 mg/g for methylene blue and 104.76 mg/g for crystal violet between pH 8–11was obtained [7].

2.2 *Chemically Modified Agro-Waste and Water Purification*

HNO_3- and NaOH-treated tangerine peel was used for the removal of heavy metal ions from aqueous solution. Approximately 89% removal occurred at pH 5.0 with 300 mg dose of adsorbent in less than 20 min (Fig. 6) shows that tangerine has highest adsorption capacity for Cu metal ions and lowest for Cd metal ions [3].

Sugarcane bagasse has been reported as a potential adsorbent for metal ions. Phosphoric acid-treated sugarcane bagasse column has been used for Pb^{2+} removal in the presence of other metal ions given (Fig. 7) [35].

A group of researchers reported the use of partially esterified sugarcane bagasse and polyacryl nitrile (PAN)-coated bagasse for the removal of various types of oil from sea water [2]. The results were compared with raw bagasse (Fig. 8). It is clearly visible from Fig. 9 that adsorption capacity improved drastically on modification [14].

Amphiphilic konjac glucomannan–chitosan (AP-KGM/CS) aerogel is reported for the removal of metal ions and dyes [22]. AP-KGM/CS is a uniquely designed material which possesses both anionic and cationic site, good specific surface area, abundant functional group, and high porosity (Fig. 9). It showed an adsorption capacity of 318 mg/g forPb^{2+}, 184 mg/g for Cu^{2+} and 94 mg/g for Cd^{2+} in mixed solution (Fig. 10) [22].

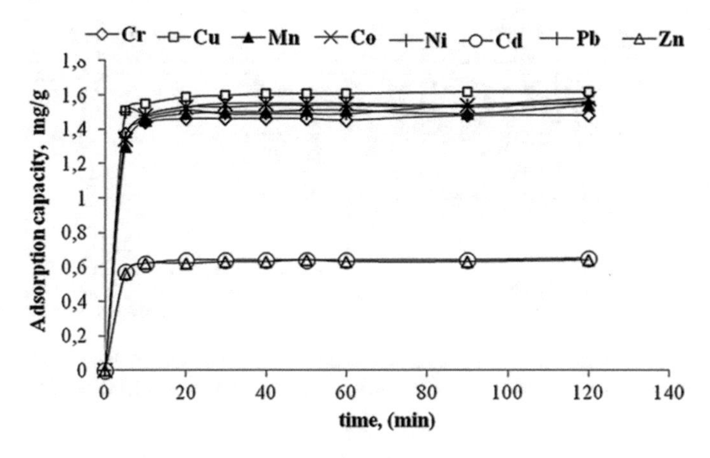

Fig. 6 Adsorption of different metal ions on tangerine peels as a function of time [3]

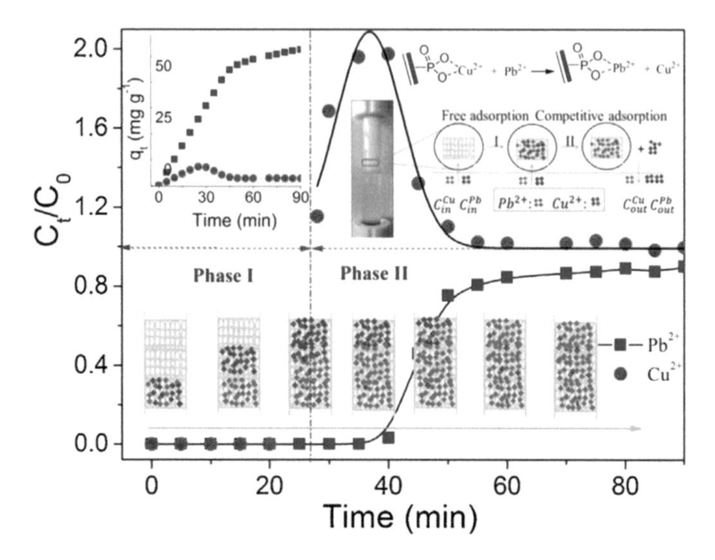

Fig. 7 Adsorption of Pb^{2+} in the presence of Cu^{2+} on phosphoric acid-treated sugarcane bagasse in fixed bed column [35]

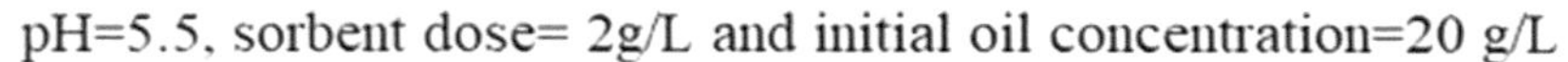

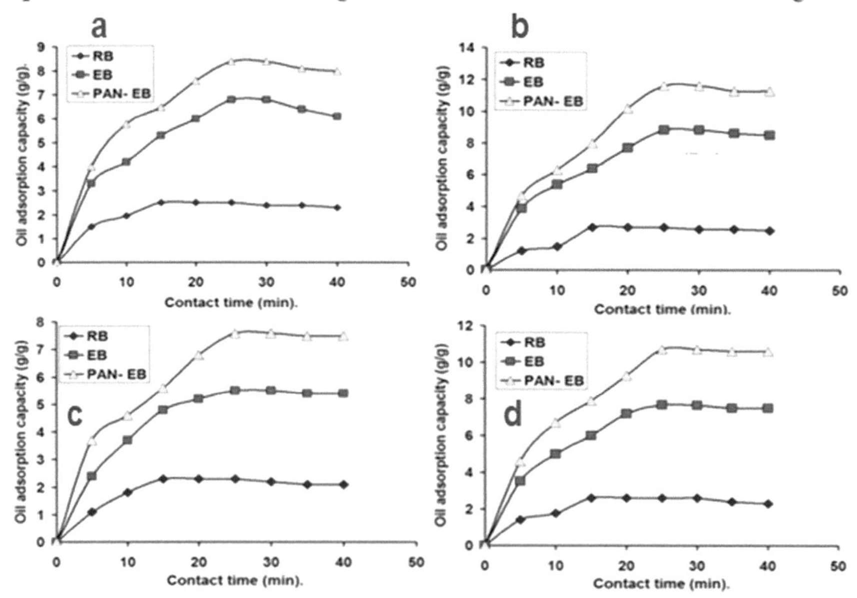

Fig. 8 Adsorption capacity of raw bagasse, esterified bagasse, and polyacryl nitrile-coated esterified bagasse for **a** diesel oil, **b** paraffin oil, **c** gasoline oil, and **d** vegetable oil [2]

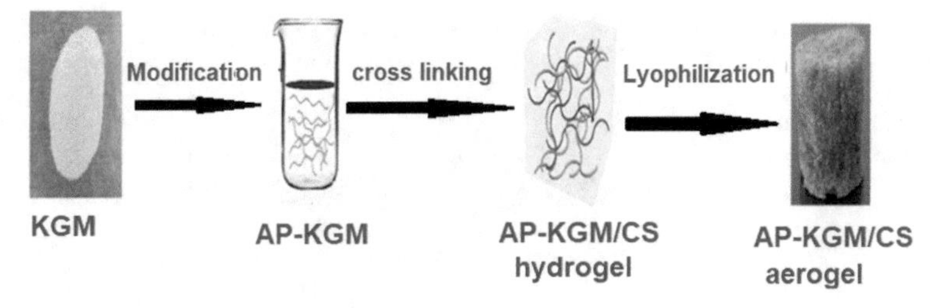

Fig. 9 Preparation scheme of amphiphilic konjac glucomannan–chitosan (AP-KGM/CS) aerogels [22]

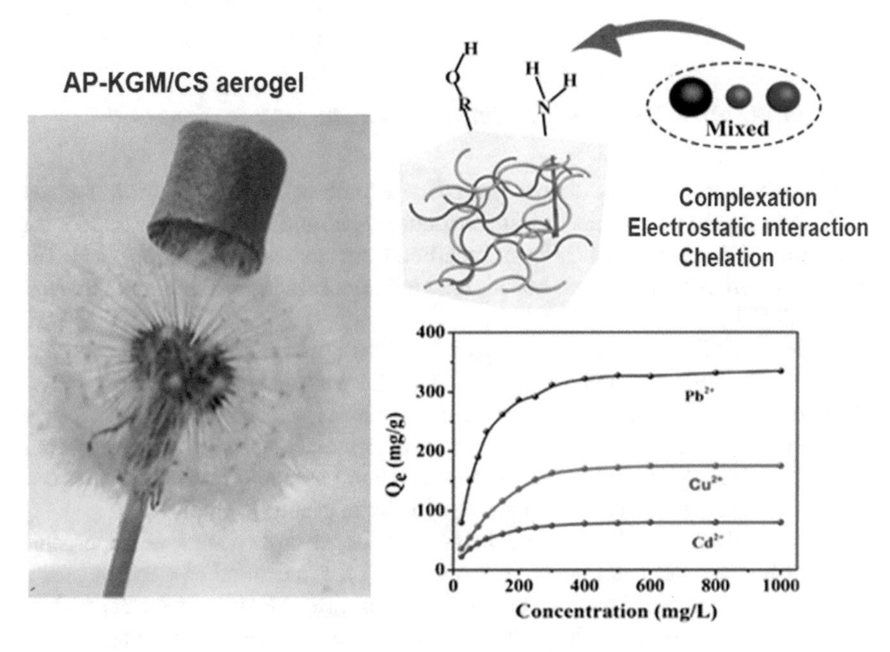

Fig. 10 Adsorption of metal ions on amphiphilic konjac glucomannan–chitosan (AP-KGM/CS) aerogels [22]

NaOH-treated rice husk is a potential adsorbent for reactive yellow dye removal [27]. Pyromellitic dianhydride (PMDA)-treated sugarcane bagasse is found to be a good adsorbent for the removal of methylene blue and rhodamine B dye from binary solution. It was noticed from breakthrough plot (Fig. 11) that after a point methylene blue substitutes the rhodamine B from active sites and get adsorbed, which proved that PMDA-modified sugarcane bagasse is better adsorbent for methylene blue dye [38].

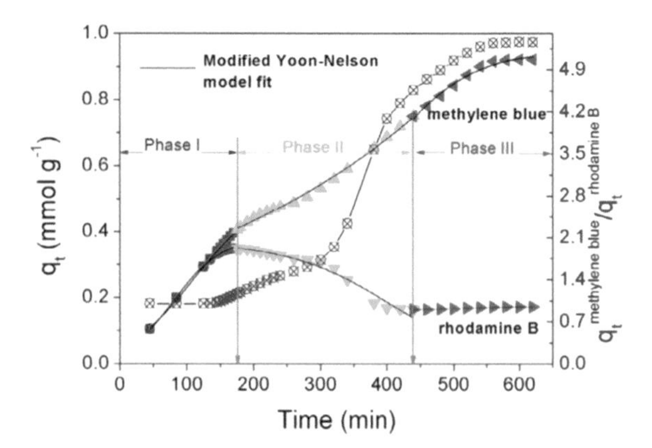

Fig. 11 Competitive adsorption of MB and Rh B on PMDA-modified sugarcane bagasse under continuous model [38]

2.3 Thermally Modified Agriculture Waste

A group of researchers from Brazil reported the use of biochar prepared from elephant grass (Pennisetum purpureum), waste coconut, guava, and orange biomasses for the adsorption of polycyclic aromatic hydrocarbon from aqueous solution (Table1). The results showed that the biomass containing high lignocellulosic content was the best adsorbent [10].

Carbonized (CO_2 atmosphere) and pyrolyzed (N_2 atmosphere) product of broccoli stalks, cauliflower cores, and coconut shell have been used as adsorbents for removal of heavy metals from aqueous solution. Biosorbent is prepared by the pyrolysis at 600 °C for 2 h acted as the best adsorbent for heavy metal ions removal from aqueous shown [21]. It is also reported that the adsorption of metal ions on the surface is due to the interaction with oxygen containing functional groups [21].

Carbonized coconut leaves have been used for adsorptive removal of organic pollutant maleic acid from water. Three methods for activation of coconut leaves were used, and the SEM pictures are given Fig. 12 [18]. SEM images clearly show the pores on the surface which provides the active sites for adsorption of maleic acid. The adsorbent prepared by chemical as well as thermal activation was found best for

Table 1 Comparison of the adsorption capacities of biochar for the polycyclic aromatic hydrocarbons in mixed solutions [10]

PAH	Orange waste biochar		Elephant waste biochar		Guava waste biochar		Coconut waste biochar	
	q_{ad}	%	q_{ad}	%	q_{ad}	%	q_{ad}	%
Benzo(a)anthracene	18.55	76.42	19.41	79.97	15.52	79.81	18.80	81.73
Benzo(b)fluoranthene	15.66	66.92	17.25	73.75	10.88	54.69	18.09	78.51
Benzo(k)fluoranthene	17.25	68.78	18.88	75.29	17.11	77.28	21.01	81.94
Dibenzo(a,h)anthracene	6.52	27.98	14.11	60.57	0.01	0.05	7.86	33.84

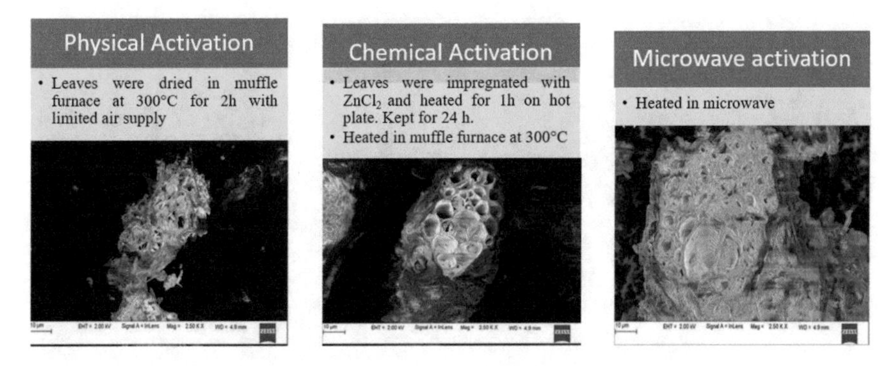

Fig. 12 Different methods for the preparation of adsorbent using dried coconut leaves

the maleic acid adsorption. The adsorption was due to hydrogen bond formation at the surface of adsorbate [18].

Carbonized marigold flower (Fig. 13) and Kailsaw dust were found to be potential adsorbent for metal ion removal and dye removal from aqueous phase [4, 14].

A number of adsorbents like spent tea leaves and other agriculture wastes were found for an efficient removal of synthetic dye and heavy metal ions from aqueous solution by batch adsorption and packed bed column methods [31].

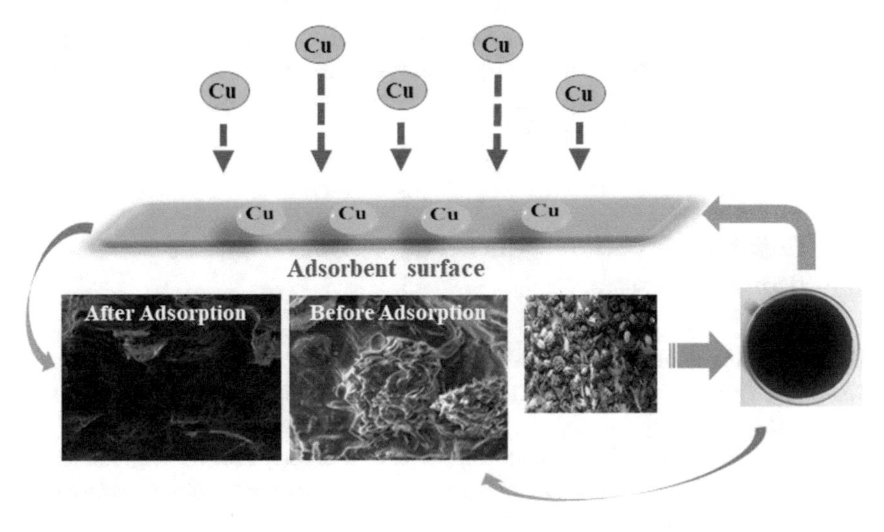

Fig. 13 Adsorption of Cu(II) metal ion on carbonized marigold flower waste [4]

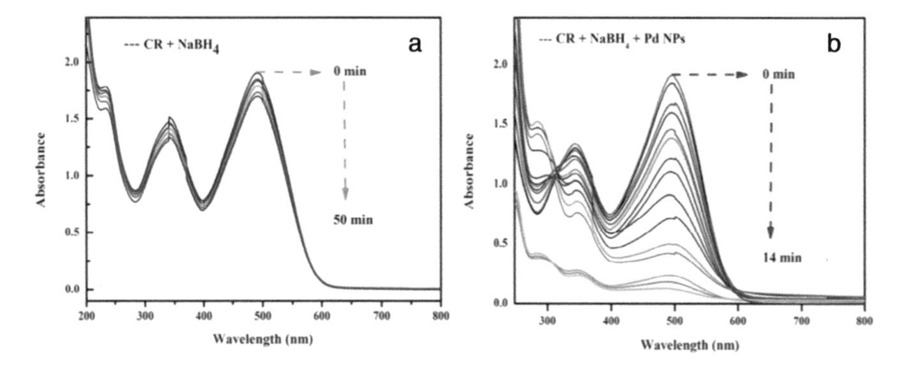

Fig. 14 Degradation of congo red dye with **a** $NaBH_4$ and **b** $NaBH_4$ + Pd nanoparticles [24]

3 Removal of Pollutants by Using Materials Derived from Agriculture Wastes

3.1 Metal and Metal Oxide Nananoparticles Obtained from Agro-Waste

A group of researchers have reported the use of cotton ball agricultural waste for Pd NPs preparation, and they have used the prepared nanoparticles catalysts to degrade azo dyes (sunset yellow, methyl orange, Tartrazine, and congo red) in combination with $NaBH_4$ from water (Fig. 14). The Pd nanoparticle catalyst showed 95–97% reduction in less than 15 min [24].

3.2 Cellulose or Cellulose-Based Composites Obtained from Agro-Waste

Nanocrystalline cellulose obtained from oil palm empty fruit bunch agricultural waste was found very efficient adsorbent for methylene blue dye removal from aqueous solution (Fig. 15) [29].

Cellulose-supported $CuInS_2$ nanocomposites are reported for the degradation of rhodamine dye (Fig. 16). Banana, orange, sweet lime, and pomegranate waste fruit rinds have been used to extract cellulose [32].

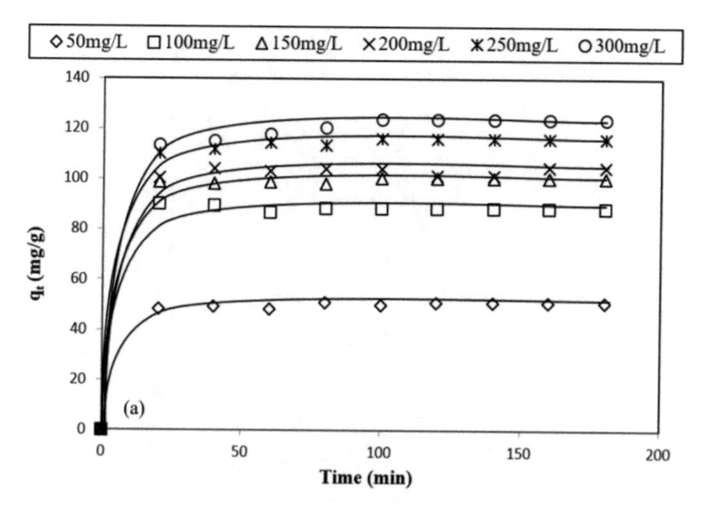

Fig. 15 Methylene blue dye adsorption on nanocrystalline cellulose [29]

3.3 Carbon Material and Carbon-Based Nanomaterials from Agro-Waste

Inorganic/organic composites derived from agro-waste

Inorganic–organic frameworks (IOFs) are emerging hybrid materials, which possess high surface area and large number of active sites. These are reported as good adsorbent for hazardous material such as dyes, insecticides, and some other organic pollutants. Synthesis of Cu-BTC@Cotton composite by coordination interaction between Cu and cellulose has been reported (Fig. 17). The prepared composite showed efficient adsorption of ethion insecticide in aqueous solution [1].

Cellulose obtained from sugarcane bagasse has been used as an adsorbent. The modification is done in cellulose by converting into magnetic biochar and doping it with N or S (Fig. 18). The prepared adsorbent is found to remove approximately 99% bisphenol A ($C_0 = 100$ ppm, $T = 343$ K) at pH 7 [5].

Biochar derived from agriculture feedstock contains carbonized and non-carbonized fraction which work as effective inorganic/organic framework for cost-effective removal of organic pollutants (Fig. 19) from water [19].

A group of researchers has reported an effective Inorganic framework by the green synthesis of biomass-derived activated carbon/Fe-Zn bimetallic nanoparticles from lemon wastes for heterogeneous Fenton-like decolorization of Reactive Red 2 dye (Fig. 20) [25].

Zhao et al. [39] has reported the use of biomass-based iron carbide nanocomposite for the removal of methyl orange dye. The prepared adsorbent is highly efficient and stable at variable pH. The composite (Fe_3C) had a core–shell structure in which zerovalent iron is coated with iron carbide and mixed in biochar. High surface area and pore volume make it effective for adsorption of methyl orange [39]. A group of

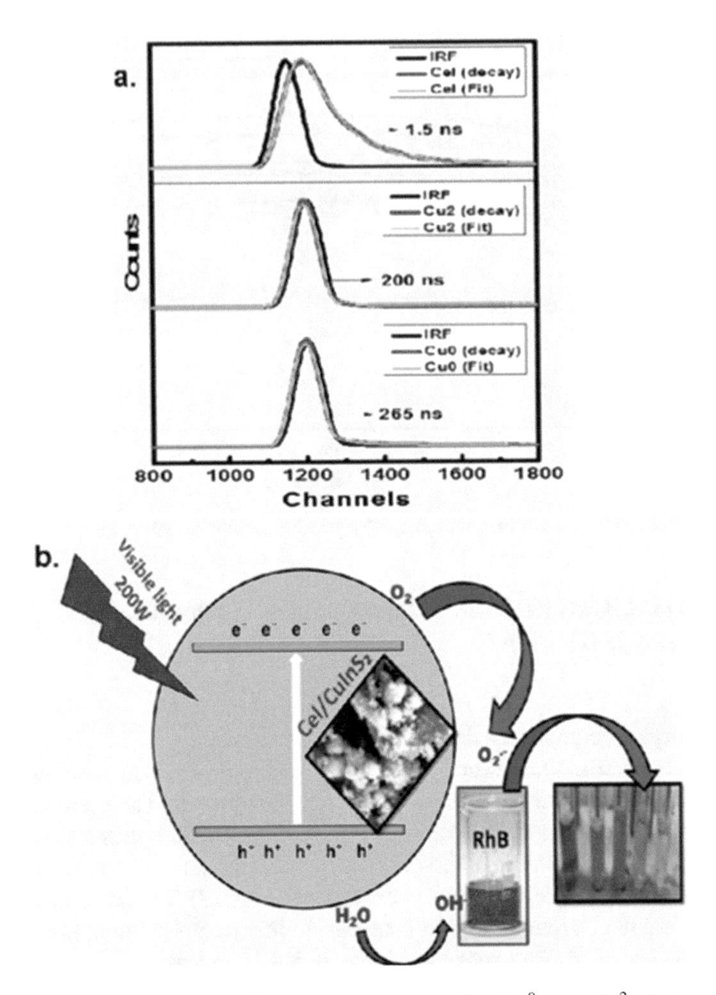

Fig. 16 **a** Time-resolved photoluminescence spectra for Cel, Cu^0, and Cu^2 photocatalysts. **b** Schematic representation of the mechanism of dye degradation process [32]

researchers has reported that amino-functionalized biomass-derived porous carbons, an organic–inorganic composite, shows enhanced aqueous adsorption affinity and sensitivity for sulfonamide antibiotics (Fig. 21) [34].

Some researchers have successfully prepared amino-functionalized biomass fly ash (BFA−APTES), an organic–inorganic composite, and performed adsorption experiment for the removal of two anionic dyes, namely alizarin red S (ARS) and bromothymol blue (BTB) [11].

Vyas et al. [33] have reported the use of Tinospora cordifolia-derived biomass functionalized ZnO particles as an organic–inorganic composite for effective removal of lead(II), iron(III), phosphate and arsenic(III) from water. Prepared adsorbent showed excellent adsorption properties. Magnetic biochar prepared using coconut

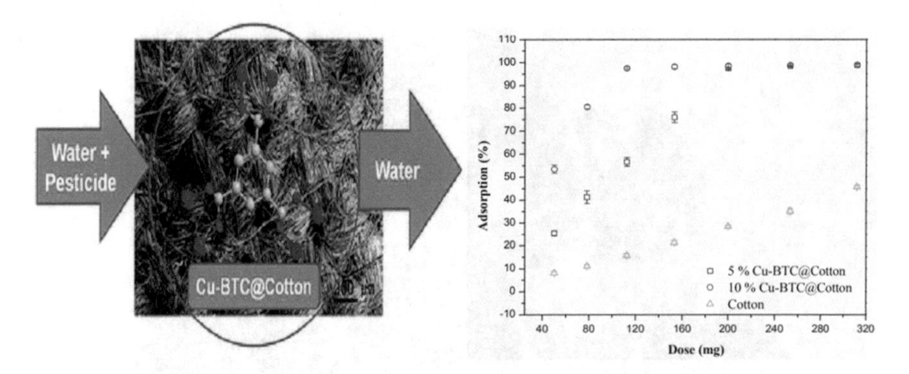

Fig. 17 **a** Adsorption of pesticide on Cu-BTC@cotton composite and **b** effect of the Cu-BTC@Cotton dose on ethion insecticide adsorption [1]

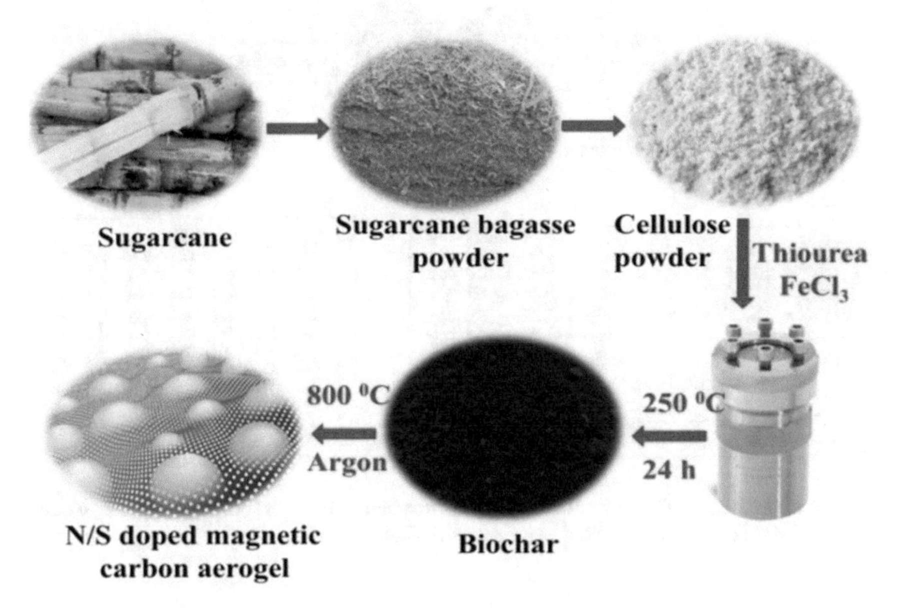

Fig. 18 Preparation of N/S-doped magnetic aerogel from sugarcane bagasse [5]

shell was used effectively for the removal of lead and cadmium from wastewater. The value of adsorption equilibrium Q_e was found 99.91 mg/g for Pb metal ions and 109.29 mg/g for Cd metal ions [36].

Sugarcane bagasse (SB), rice straw (RS), peanut shells (PS), and herb residue (HR) were pulverized, sieved, and immersed in diluted steel pickling waste liquor which contain approximately 12 g/L of iron to make the biomass iron rich. The obtained material was then heated in muffle furnace, meshed, and sieved. The prepared magnetic biochars (BC) were used for the removal of Cr (VI) metal ions. The effect of zeta potential on the removal of Cr (VI) has also been investigated. The

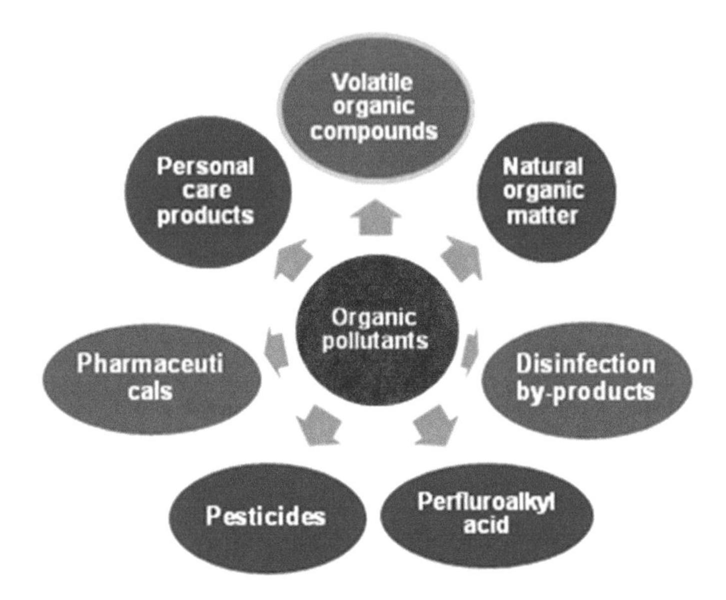

Fig. 19 Types of organic pollutants

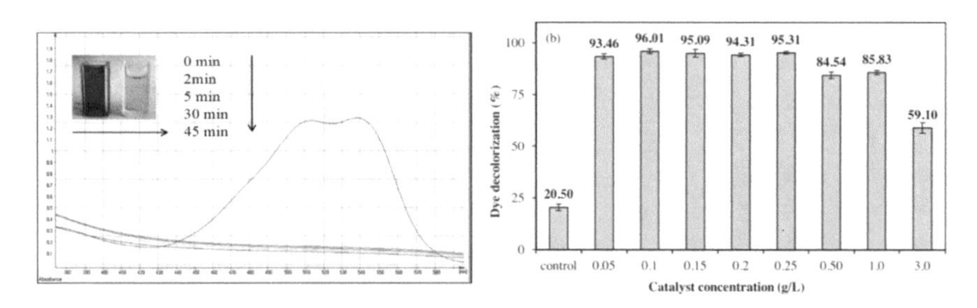

Fig. 20 **a** Monitoring of Reactive Red 2 dye decolorization by UV–Vis spectrophotometer. **b** effect of catalyst dose on the Reactive Red 2 dye decolorization [25]

magnetic biochar shows the positive zeta potential than the raw biochar (Fig. 22). It is observed that higher the positive charge on magnetic biochar, higher the removal of Cr (VI) metal ion [37].

Iron oxide and CuO nanoparticles fabricated rice husk biochar have been reported for the removal of arsenic and microbial contamination from water. The effect of pH and SO_4^-, PO_4^-, and F^- ions on the removal arsenic at pH 7.0 was found more appropriate for arsenic removal. Increasing concentration of SO_4^-, PO_4^-, and F^- ions showed negative impact on adsorption of arsenic [26]. Ethylene diamine/Fe(III)-functionalized saw dust is an effective and recyclable phosphate-binding ligand (39.9 $\pm$ 8.1 g/kg) in agricultural runoff (Fig. 23). The recovery of phosphate is also possible by the adsorbent [23].

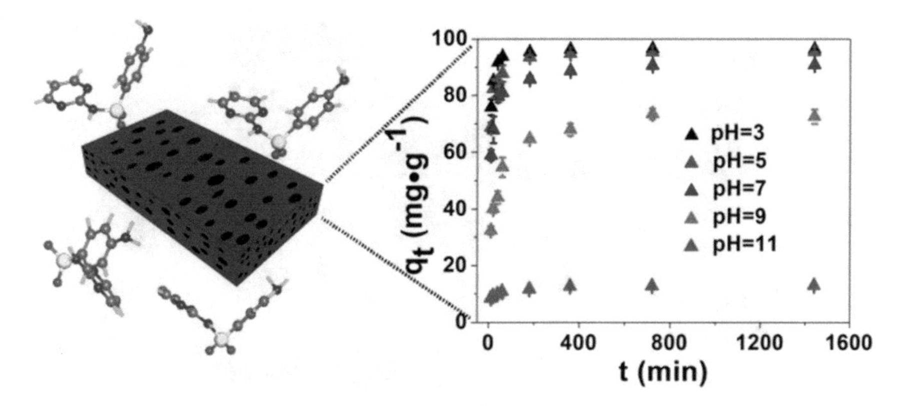

Fig. 21 Adsorption of sulfonamide antibiotics on amino-functionalized biomass-derived porous carbon [34]

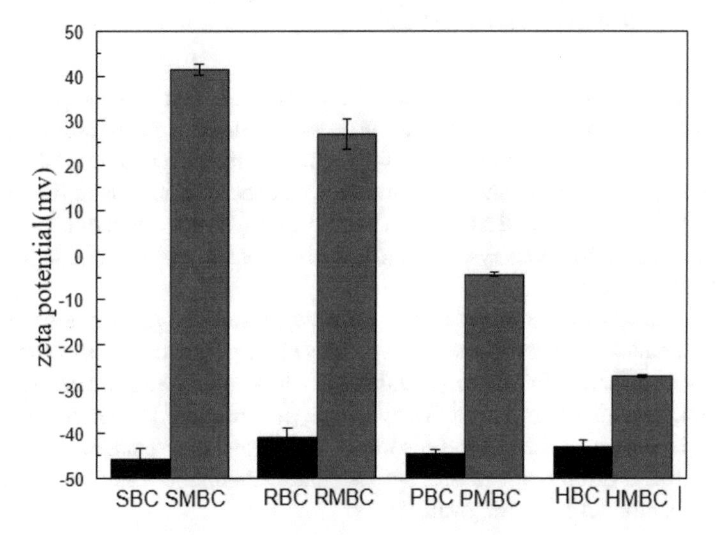

Fig. 22 Zeta potential of magnetic biochar obtained from sugarcane bagasse, rice straw, peanut shells, and herb residue (HR) [37]

Magnetic composite polyethyleneimine with sugarcane bagasse (Fe_3O_4@PEI/SCB) was prepared and analyzed by SEM, XRD, and FTIR. It was used for the removal of Cr (VI) and Orange II dye from aqueous solution. The Langmuir model was found to fit the data [12].

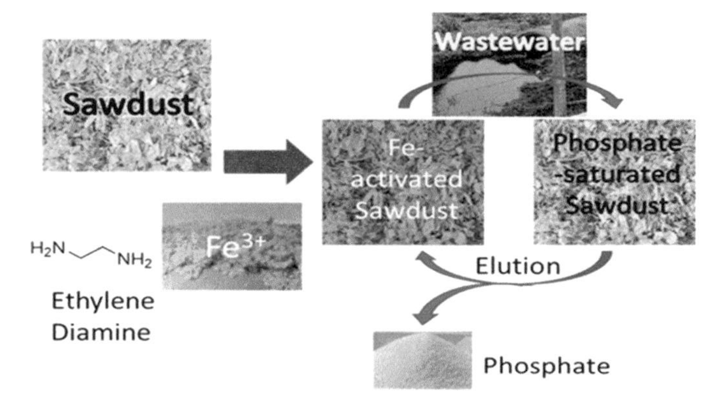

Fig. 23 Phosphate removal from ethylene diamine/Fe(III)-functionalized sawdust [23]

4 Adsorption Isotherm Models

The adsorption isotherms are helpful for the determination of adsorption capacities of used adsorbents. In adsorption process, when the concentration of adsorbate molecules in the solution is in equilibrium with the concentration of adsorbate molecules on the surface of adsorbent, a state is reached at which adsorption becomes constant. This stage is called adsorption equilibrium. The condition of equilibrium is a characteristic of the entire system, e.g., adsorbate, adsorbent, solvent, temperature, pH, particle size, etc.

The equilibrium data at a fixed temperature is represented by adsorption isotherms. The study of adsorption isotherms gives an idea of adsorbent capacity for adsorption of specific adsorbate. Monolayer or multilayer surface phase may be involved in the adsorption. Several adsorption isotherm models were proposed to determine the ratio of dye molecules, adsorbed on the adsorbent surface and that remained in solution at equilibrium at a given temperature, but Langmuir, Freundlich, and Temkin models are widely studied for dye removal.

4.1 *Langmuir Isotherm Model*

The Langmuir isotherm model explains the behavior of adsorbate as an ideal gas at isothermal condition. It predicts homogeneous monolayer adsorption of adsorbates on equivalent adsorbent sites. The linearized equation for the model can be written as:

$$\frac{C_e}{q_e} = \frac{1}{Q^0 b} + \frac{C_e}{Q^0} \tag{1}$$

where C_e and q_e are the remaining concentration (mg/L) and amount of dye adsorbed (mg/g) at equilibrium, respectively, while Q^0 (mg/g) and b (L/mg) both are the Langmuir constants. This model is failed to account for the roughness of surface. It also neglects the adsorbate–adsorbent interaction on nearby sites.

4.2 Freundlich Isotherm Model

Freundlich isotherm model explains the relation between quantities of adsorbate adsorbed on the adsorbent surface to the remaining quantity of adsorbate in the contacting solution. It predicts multisite adsorption on the heterogeneous surface. The linearized equation can be written as:

$$\log q_e = \log K_f + \frac{1}{n} \log C_e \tag{2}$$

where K_f and n both are Freundlich constants, used for determining adsorption capacity (L/mg) and adsorption intensity, respectively. This model is not suitable at high pressure.

4.3 Temkin Isotherm Model

The Temkin isotherm model explains the indirect adsorbate–adsorbent interaction on nearby active sites. The linearized equation for Temkin adsorption isotherm is given by following Eq. 3.

$$\mathrm{q_e = B\ lnA + B\ lnC_e} \tag{3}$$

where $B = \frac{RT}{b}$.

A (*L*/*g*) and *B* are Temkin constants, *R* is the gas constant, *T* is the temperature in Kelvin, and *b* (J/mol) is a constant connected to the heat of adsorption.

5 Kinetic Models

The kinetics of adsorption can be understood by considering following two kinetic models. Integrated pseudo-first-order and pseudo-second-order rate expressions are given by Eqs. 4 and 5, respectively.

$$\log(q_e - q) = \log\ q_e - \frac{k_1}{2.303}t \tag{4}$$

$$\frac{t}{q} = \frac{1}{k_2\ q_e^2} + \frac{t}{q_e} \tag{5}$$

where $q_e$and q are quantities of adsorbates, adsorbed at equilibrium and at various time t, k_1 and $k_2$are the pseudo-first-order and pseudo-second-order rate constants.

The plots of log (q_e–q) versus t for a pseudo-first order and t/q versus t for pseudo-second order are used to determine the kinetic constants. The suitability of the kinetic model for a particular adsorbate–adsorbent system is decided by the values of correlation coefficients (R_1^2 and R_2^2).

6 Mechanism of Adsorption

The sorption of organic pollutants on biochar takes place by different kind of interactions (Fig. 24) [19].

The mechanism of decolorization of Reactive Red dye by biomass-derived activated carbon/Fe-Zn bimetallic nanoparticles is given below [25].

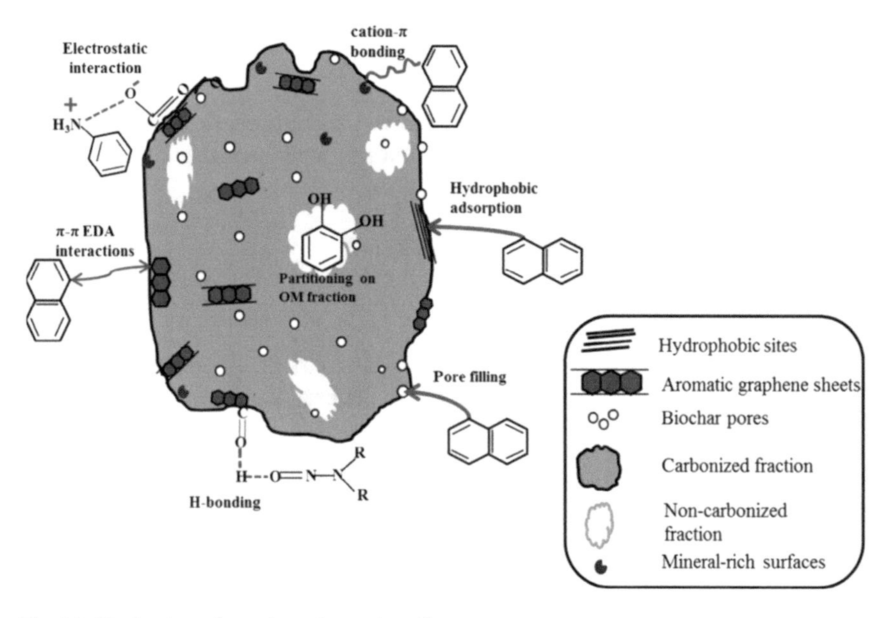

Fig. 24 Mechanism of sorption of organic pollutants

$$S - Fe^{2+} + H_2O_2 \rightarrow OH^{\cdot} + OH^- + S - Fe^{3+}$$
$$S - Fe^{3+} + H_2O_2 \rightarrow S - Fe^{2+} + H^+ + HO_2^{\cdot}$$
$$S - Fe^{3+} + HO_2^{\cdot} \rightarrow S - Fe^{2+} + H^+ + O_2$$

$$S - Fe^{2+} + OH^{\cdot} \rightarrow S - Fe^{3+} + OH^-$$
$$OH^{\cdot} + OH^{\cdot} \rightarrow H_2O_2$$
$$OH^{\cdot} + H_2O_2 \rightarrow H_2O + HO_2^{\cdot}$$
$$OH^{\cdot} + \text{organic molecules} \rightarrow CO_2 + H_2O$$

Degradation mechanism of methyl orange dye Fe^0 and Fe_3C composite is given in Fig. 25 [39].

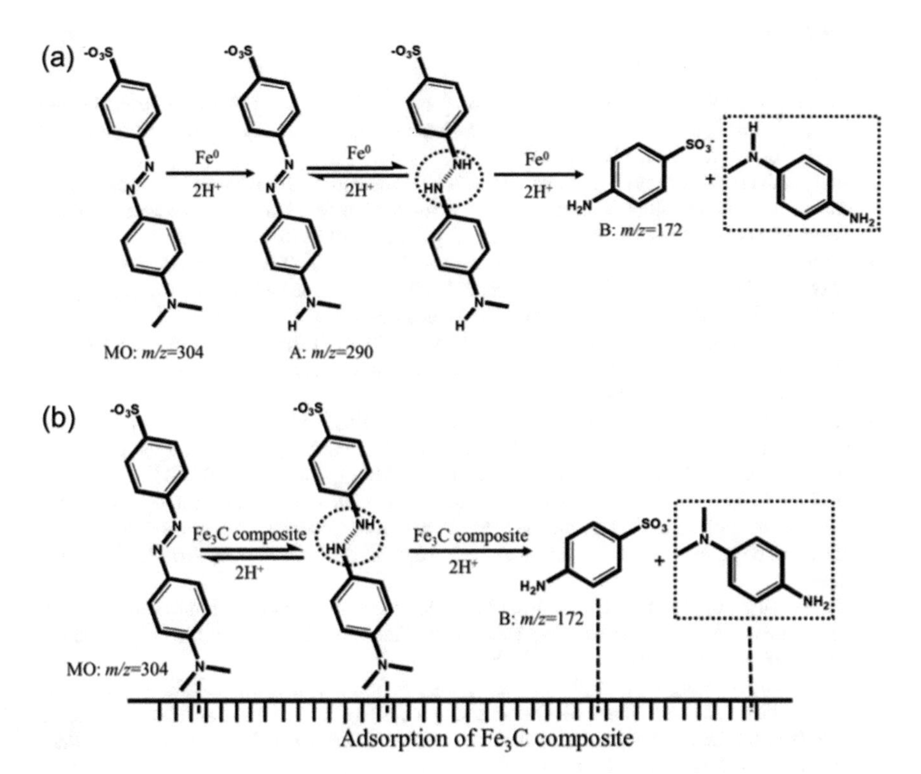

Fig. 25 Proposed degradation pathway of MO by Fe^0(**a**) and Fe_3C composite (**b**) dotted square means the product was not detected in the solution, which might be adsorbed on the surface of Fe^0 and Fe_3C composite [39]

7 Conclusions

Agro-industrial wastes (raw and modified form) have been used as adsorbents for the removal of different type of pollutants from water. It is reported that the modified form or derived products having higher surface area and porosity are better adsorbents and are being used frequently. In this chapter, different agricultural wastes and derived or modified products have been discussed for decontamination of different pollutants from water. Effect of different parameters on removal efficiency has been described. Adsorption isotherm models and kinetic models have been discussed. Process of desorption and reuse is also pointed out.

References

1. Abdelhameed RM, Abdel-Gawad H, Elshahat M, Emam HE (2016) Cu–BTC@cotton composite: design and removal of ethion insecticide from water. RSC Adv 6:42324–42333. https://doi.org/10.1039/C6RA04719J
2. Abdelwahab NA, Shukry N, El-kalyoubi SF (2017) Preparation and characterization of polymer coated partially esterified sugarcane bagasse for separation of oil from seawater. Environ Technol 38(15):1905–1914. https://doi.org/10.1080/09593330.2016.1240243
3. Abdic S, Memic M, Sabanovic E, Sulejmanovic J, BegicS (2018) Adsorptive removal of eight heavy metals from aqueous solution by unmodified and modified agricultural waste: Tangerine peel. Int J Environ Sci Technol 15:2511–2518. https://doi.org/10.1007/s13762-018-1645-7
4. Agarwal A, Kumar A, Gupta P, Tomar R, Singh NB (2020) Cu(II) ions removal from water by charcoal obtained from marigold flower waste. Materials Today: Proceedings. https://doi.org/10.1016/j.matpr.2020.11.046
5. Ahamad T, Naushad M, Ruksana AAN, Alshehri SM (2019) N/S doped highly porous magnetic carbon aerogel derived from sugarcane bagasse cellulose for the removal of bisphenol-A. Int J Biol Macromol 132:1031–1038. https://doi.org/10.1016/j.ijbiomac.2019.04.004
6. Ansari SA, Khan F, Ahmad A (2016) Cauliflower leave, an agricultural waste biomass adsorbent, and its application for the removal of MB dye from aqueous solution: equilibrium, kinetics, and thermodynamic studies. Int J Anal Chem 2016:8252354. https://doi.org/10.1155/2016/8252354
7. Bhattacharjee C, Dutta S, Saxena VK (2020) A review on biosorptive removal of dyes and heavy metals from wastewater using watermelon rind as biosorbent. Environ Adv 2:100007. https://doi.org/10.1016/j.envadv.2020.100007
8. Dai Y, Sun Q, Wang W, Lu L, Liu M, Li J, Yang S, Sun Y, Zhang K, Xu J, Zheng W, Hu Z, Yang Y, Gao Y, Chen Y, Zhang X, Gao F, Zhang Y (2018) Utilizations of agricultural waste as adsorbent for the removal of contaminants: a review. Chemosphere 211:235–253. https://doi.org/10.1016/j.chemosphere.2018.06.179
9. Dalibor S, Eva P (2017) Sorption of iron, manganese, and copper from aqueous solution using orange peel: optimization, isothermic, kinetic, and thermodynamic studies. Pol J Environ Stud 26(2):795–800. https://doi.org/10.15244/pjoes/60499
10. De Jesus JHF, da Matos TT, da Cunha G, Mangrich AS, Romao LPC (2019) Adsorption of aromatic compounds by biochar: influence of the type of tropical biomass precursor. Cellulose 26:4291–4299. https://doi.org/10.1007/s10570-019-02394-0
11. Dogar S, Nayab S, Farooq MQ, Said A, Kamran R, Duran H, Yameen B (2020) Utilization of biomass fly ash for improving quality of organic dye-contaminated water. ACS Omega 5(26):15850–15864. https://doi.org/10.1021/acsomega.0c00889

12. Gao H, Du J, Liao Y (2019) Removal of chromium(VI) and orange II from aqueous solution using magnetic polyetherimide/sugarcanebagasse. Cellulose 26:3285–3297. https://doi.org/10.1007/s10570-019-02301-7
13. Goodman BA (2020) Utilization of waste straw and husks from rice production: a review. J Bioresour Bioprod 5(3):143–162. https://doi.org/10.1016/j.jobab.2020.07.001
14. Gupta V, Agarwal A, Singh MK, Singh NB (2019) Kail sawdust charcoal: a low-cost adsorbent for removal of textile dyes from aqueous solution. SN Appl Sci 1:1271. https://doi.org/10.1007/s42452-019-1252-3
15. Gusain R, Kumar N, Ray SS (2020) Recent advances in carbon nanomaterial-based adsorbents for water purification. Coord Chem Rev 405:213111. https://doi.org/10.1016/j.ccr.2019.213111
16. Hameed B, Hakimi H (2008) Utilization of durian (Duriozibethinus Murray) peel as low cost sorbent for the removal of acid dye from aqueous solutions. Biochem Eng J 39(2):338–343. https://doi.org/10.1016/j.bej.2007.10.005
17. Hayder AA, Mohammad NA, Zaidun NA, Alaa HA (2018) Adsorption of thallium ion (Tl^{+3}) from aqueous solutions by rice husk in a fixed-bed column: Experiment and prediction of breakthrough curves. Environ Technol Innov 12:1–13. https://doi.org/10.1016/j.eti.2018.07.001
18. Hemashree K, Bhat JI (2017). Synthesis, characterization and adsorption behavior of coconut leaf carbon. Res Chem Intermed 43(8):4369–4386. https://doi.org/10.1007/s11164-017-2883-x
19. Inyang M, Dickenson E (2015) The potential role of biochar in the removal of organic and microbial contaminants from potable and reuse water: a review. Chemosphere 134:232–240. https://doi.org/10.1016/j.chemosphere.2015.03.072
20. Islamuddin G, Khalid MA, Ahmad SA (2019) Study of eco-friendly agricultural wastes as non-conventional low cost adsorbents: a review. Ukrainian J Ecol 9:68–75
21. Landin-Sandoval VJ, Mendoza-Castillo DI, Bonilla-Petriciolet A, Aguayo-Villarreal IA, Reynel-Avila HE, Gonzalez-Ponce HA (2020) Valorization of agri-food industry wastes to prepare adsorbents for heavy metal removal from water. J Environ Chem Eng 8(5):104067. https://doi.org/10.1016/j.jece.2020.104067
22. Mao J, Li S, He C (2019) Robust amphiprotic konjac glucomannan cross-linked chitosan aerogels for efficient water remediation. Cellulose 26:6785–7679. https://doi.org/10.1007/s10570-019-02549-z
23. Meister D, Ure D, Awada A, Barrette JC, Gagnon J, Mutus B, Trant JF (2019) Covalently Functionalized Sawdust for the Remediationof Phosphate from Agricultural Wastewater. ACS Sustain Chem Eng 7(24):20139–20150. https://doi.org/10.1021/acssuschemeng.9b06073
24. Narasaiah BP, Mandal BK (2020) Remediation of azo-dyes based toxicity by agro-waste cotton boll peels mediated palladium nanoparticles. J Saudi Chem Soc 24(2):267–281. https://doi.org/10.1016/j.jscs.2019.11.003
25. Oruç Z, Ergüt M, Uzunoğlu D, Özer A (2019) Green synthesis of biomass-derived activated carbon/Fe-Zn bimetallic nanoparticles from lemon (*Citrus limon* (L.) Burm. f.) wastes for heterogeneous Fenton-like decolorization of Reactive Red 2. J Environ Chem Eng 7(4):103231. https://doi.org/10.1016/j.jece.2019.103231
26. Priyadarshni N, Nath P, Nagahanumaiah CN (2020) Sustainable removal of arsenate, arsenite and bacterial contamination from water using biochar stabilized iron and copper oxide nanoparticles and associated mechanism of the remediation process. J Water Process Eng 37:101495. https://doi.org/10.1016/j.jwpe.2020.101495
27. Rachna K, Agarwal A, Singh NB (2019) Rice husk and Sodium hydroxide activated Rice husk for removal of Reactive yellow dye from water. Mater Today Proc 12(3):573–580. https://doi.org/10.1016/j.matpr.2019.03.100
28. Shakoor S, Nadar A (2017) Adsorptive treatment of hazardous methylene blue dye from artificially contaminated water using cucumis sativus peel waste as a low-cost adsorbent. Groundwater Sustain Develop 5:152–159. https://doi.org/10.1016/j.gsd.2017.06.005

29. Shanmugarajah B, Chew IM, Mubarak NM, Choong TSY, Yoo CK, Tan KW (2019) Valorization of palm oil agro-waste into cellulose biosorbents for highly effective textile effluent remediation. J Clean Prod 210:697–709. https://doi.org/10.1016/j.jclepro.2018.10.342
30. Singh NB, Nagpal G, Agrawal S, Rachna (2018) Water purification by using adsorbents: a review. Environ Technol Innov 11:187–240. https://doi.org/10.1016/j.eti.2018.05.006
31. Sulyman M, Namiesnik J, Gierak A (2017) Low-cost adsorbents derived from agricultural by-products/wastes for enhancing contaminant uptakes from wastewater: a review. Polish J Environ Stud 26(2):479–510. https://doi.org/10.15244/pjoes/66769
32. Tavker N, Gaur UK, Sharma M (2019) Highly active agro-waste-extracted cellulose-supported $CuInS_2$ nanocomposite for visible-light-induced photocatalysis. ACS Omega 4 (7):11777–11784. https://doi.org/10.1021/acsomega.9b01054
33. Vyas G, Bhatt S, Paul P (2019) Tinospora cordifolia derived biomass functionalized ZnO particles for effective removal of lead(II), iron(III), phosphate and arsenic(III) from water. RSC Adv 9:34102. https://doi.org/10.1039/C9RA07042G
34. Wang Y, Jiao WB, Wang JT, Liu GF, Cao HL, Lü J (2019) Amino-functionalized biomass-derived porous carbons with enhanced aqueous adsorption affinity and sensitivity of sulfonamideantibiotics. Biores Technol 277:128–135. https://doi.org/10.1016/j.biortech.2019.01.033
35. Xiong W, Zhang J, Yu J, Chi R (2019) Competitive adsorption behavior and mechanism for Pb^{2+} selective removal from aqueous solution on phosphoric acid modified sugarcane bagasse fixed-bed column. Process Saf Environ Prot 124:75–83. https://doi.org/10.1016/j.psep.2019.02.001
36. Yap MW, Mubarak NM, Sahu JN, Abdullah EC (2017) Microwave induced synthesis of magnetic biochar from agricultural biomass for removal of lead and cadmium from wastewater. J Indus Eng Chem 45:287–295. https://doi.org/10.1016/j.jiec.2016.09.036
37. Yi Y, Tu G, Zhao D, Tsang PE, Fang Z (2018) Biomass waste components significantly influence the removal of Cr(VI) using magnetic biochar derived from four types of feedstocks and steel pickling waste liquor. Chem Eng J 360:212–220. https://doi.org/10.1016/j.cej.2018.11.205
38. Yu J, Zhu J, Feng L, Cai X, Zhang Y, Chi R (2015) Removal of cationic dyes by modified waste biosorbent under continuous model: competitive adsorption and kinetics. Arab J Chem 12(8):2044–2051. https://doi.org/10.1016/j.arabjc.2014.12.022
39. Zhao N, Chang F, Hao B, Yu L, Morel JL, Zhang J (2019) Removal of organic dye by biomass-based iron carbide composite with an improved stability and efficiency. J Hazard Mater 5(369):621–631. https://doi.org/10.1016/j.jhazmat.2019.02.077

Mesoporous Materials for Adsorption of Heavy Metals from Wastewater

Nabila Bensacia, Ioana Fechete, Khalida Boutemak, and Ahmed Kettab

Abstract Heavy metals are substances which present a grave danger for human beings and ecosystems, having impacts on plants, consumer products, and humans. In this topic, we will focus on the treatment of wastewater using the adsorption technique, specifically reviewing; (1) the general properties of heavy metals; (2) the current knowledge on the adsorption process; (3) how the chemical modification of the surface of the mesoporous materials makes a great possibility to create new structures, which have new functional groups, making the interaction with heavy metals better in comparison to the basic structures, therefore obtaining new high quality adsorbents.

Keywords Adsorption · Mesoporous sorbents · Heavy metals · Functionalization · Porous silica · Wastewater · Grafting Methods · Post-synthesis grafting · Co-condensation · Contamination

Abbreviations

MCM-n Mobil composition of matter (ordered mesoporous silica material)

N. Bensacia (✉)
Hydrogen Energy Application Laboratory, Department of Process Engineering, Faculty of Technology, University of Blida1, Street of Soumaa pb 270 , Blida, Algeria

I. Fechete
ICD-LASMIS, Antenne de Nogent, Nogent International Center for CVD Innovation-NICCI, LRC-CEA-ICD-LASMIS Sud Champagne Technology Center, University of Technology Troyes, 26 street. Lavoisier, 52800 Nogent, France
e-mail: ioana.fechete@utt.fr

K. Boutemak
Functional Analysis of Chemical Processes Laboratory, Department of Process Engineering, Faculty of Technology, University of Blida1, Street of Soumaa pb 270 , Blida, Algeria

A. Kettab
National Polytechnic School, University of Bouira–Algeria, Bouira, Algiers, Algeria

E. Lichtfouse et al. (eds.), *Inorganic-Organic Composites for Water and Wastewater Treatment*, Environmental Footprints and Eco-design of Products and Processes,
https://doi.org/10.1007/978-981-16-5916-4_8

M41S	Family of molecular sieves of mobil (ordered mesoporous silica material)
SBA-n	Santa barbara amorphous (ordered mesoporous silica material)
KiT	Korea advanced institute of science and technology (ordered mesoporous silica material)
HMS-n	Hexagonal mesoporous silica
TUD	Ordered mesoporous silica material
FSM-n	Folder sheet mesoporous (ordered mesoporous silica material)
FDU-n	Fudan university (ordered mesoporous silica material)
MSU-n	Silica, mesostructured (ordered mesoporous silica material)
1-D, 2-D, 3-D	One-, two-, and three- dimensional

1 Introduction

As different industries had developed various new technologies came to existence, yet industrial discharges were still not treated in a suitable way. Thus attributing to various human activities and affecting; surface water, fresh water, and groundwater [8, 22]. Water pollution can affect plants, animals and aquatic areas. In general, water pollution has adverse effects not only on humans but also on the biospheres [42]. The problem of water pollution intensifying was due on one hand to the increase in the number of population and on the other hand to the increase of industrial activities.

The main source of heavy metal existence in water surfaces is the industrial activities such as: mining and manufacturing of electronic devices, the heavy metals generated by these activities are numerous, such as; copper, cobalt, zinc, nickel, lead and chromium in water. This toxicity is attributed mainly to the non-biodegradability of heavy metals and the tendency of heavy metals to accumulate in living organisms and many heavy metal ions are found to be carcinogenic elements [39, 51].

Industrial effluents and discharges containing traces of metal must be treated before being released into nature. Different approaches are applied to eliminated heavy metals from contaminated water, including ion exchange, coagulation and flocculation, membrane techniques, precipitation, advanced oxidation processes, electrochemical treatments and adsorption [1, 59].

The Adsorption technique is considered to be the most suitable approach to eliminate heavy metals or also called trace metal elements. Indeed, the adsorption phenomenon was given a facile design and treatment, and normally offers an effluent having acceptable degree of treatment. The Adsorption process is defined by the accumulate of gas or liquid solute on the surface of a solid or a liquid, forming a molecular or atomic film. However, adsorption is the phenomenon of atoms or molecules of gas or liquid adhesion to a surface and this operation creates a film of the adsorbate with interaction between adsorbent and adsorbate on the surface of the adsorbent.

Seeing as there is a considerable development regarding different activities' number of units, the quantity of industrial discharges and effluents discharged into natural environments continues to increase. Among these discharged pollutants, heavy metals or trace metal elements have gained the attention of researchers and scholars to develop new technologies [65].

Many adsorbents have been studied in the elimination of heavy metals from contaminated water [9, 52, 53], but the need to create novel and high efficiency adsorbents is growing exponentially. Mesoporous hybrid materials are promising adsorbents for elimination of heavy metals from wastewaters.

The adsorption technique employed for the removal of heavy metals using an adsorbent based on mesoporous silica is a very promising solution and this is due to the properties of these materials. Mesoporous silica can easily be chemically modified and reused [41]. Due to these properties, great importance is given to studies concerning the elimination of heavy metals by the adsorption technique using mesoporous silica as an adsorbent [43].

2 Contamination of Wastewater by Heavy Metals

Heavy metals, otherwise known as metallic trace elements, are metallic elements known by toxicity and non-biodegradability especially in aquatic environments [58]. The existence of heavy metals in wastewater is mainly due on the one hand to the production of this kind of effluent by a large number of industrial units, among these units we can mention; leather tanning, chemical manufacturing and semiconductor.

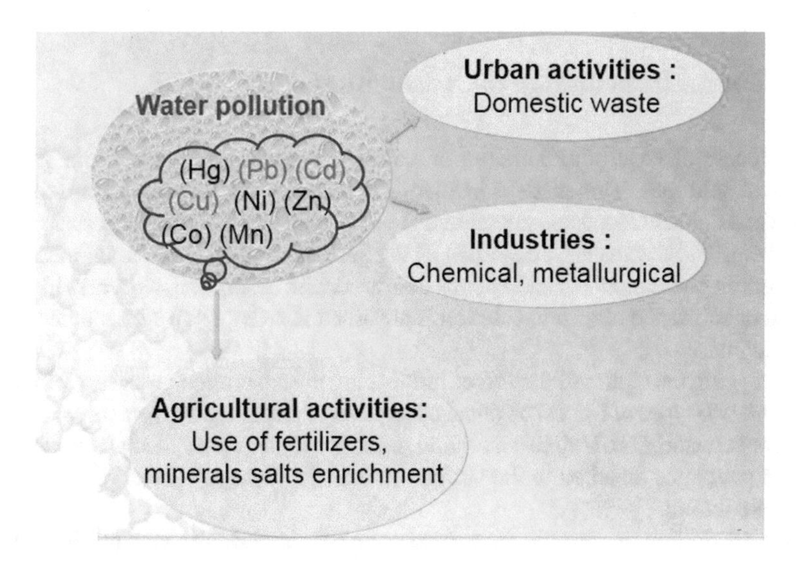

Fig. 1 Few sources of contamination of water by heavy metals

Table 1 Toxicity for some heavy metals [11, 44]

Heavy metals	Toxicity
Arsenic	Night blindness, heart sickness
Cadmium	Acute bronchopneumopathy
Chromium	Poisoning, cancer
Copper	Neurological disorders
Zinc	Neurological disorders
Lead	Poisoning, chronic renal failure
Mercury	Acute encephalopathy, Azotemic nephritis

On the other hand, they exist due to the discharge of effluents laden with heavy metals without adequate treatment in aquatic environments [63]. Figure 1 represents main sources of heavy metals entering the aquatic media.

Among the properties of heavy metals, is that they can accumulate inside the tissues of plants, animals and humans through inhalation and manual handling. This can interfere with the functioning of vital cells, causing critical situations especially for humans [25, 66]. Extensive exposure to heavy metals can cause illnesses such as: neurodegenerative sickness and cancer [10, 67]. The table 1 below shows toxicity of some heavy metals [11, 44].

Metallic substances have a high level of toxicity, which has attracted researchers' intention to develop new techniques that will be used in the elimination of heavy metals from wastewater in order to protect our planet.

3 General Information on Adsorption

The process of adsorption is based on solids to remove substances from gaseous or liquid solutions. This method involves the transfer of a substance from the first phase and demands focus on the second phase. In general, the adsorbing phase is the adsorbent, and the material adsorbed at the surface of that phase is the adsorbate. Adsorption phenomenon should not be confused with absorption, a process in which a fluid or substance of a solid solution is absorbed into the volume of another liquid or solid phase.

Analyzing the nature of the forces influencing the adsorption process, we can say there are two types of forces: physical adsorption or physisorption; chemical adsorption or chemisorption. Unlike adsorption, there is another term called desorption that means pollutants attached to the surface of the adsorbent are detached and comes back to solution.

The adsorption process can be carried out either in discontinuous/batch mode or continuous mode. The discontinuous mode is generally used on a small scale such as laboratories. However, the continuous mode is favored, and therefore used on a large scale such as industries [33]. Adsorption kinetics is directly influenced by temperature, pH, initial concentration, adsorbent size, the ionic strength [54].

Adsorption kinetics, expressed in terms of solute retention rates as a function of contact time, and is one of the most important characteristics defining the efficiency of adsorption [12, 39, 59]. The kinetic study of adsorption processes provides information on the mechanism of adsorption and the mode of transfer of solutes from the liquid phase to the solid phase.

3.1 *Pseudo-First-Order Model*

Lagergren [47] indicated that the adsorption behavior of a substance towards an adsorbent is based on adsorption capacity of adsorbent. The pseudo-first-order equation is mentioned by the following equation (Eq. 1):

$$\frac{dQ}{dt} = k_1(Q_e - Q_t) \tag{1}$$

$k1$ signifies the pseudo-first-order constant; Qe signifies the adsorption capacity of the adsorbent (mg/g) and Qt signifies the quantity of the adsorbate at time t (mg/g). The linear equation is indicated by Eq. 2:

$$\text{In}(Q_e - Q_t) = \text{In Q}_e - k_1{}^t/2.3030 \tag{2}$$

3.2 *Pseudo-Second-Order Model*

The pseudo-second-order model is influenced by the number of active surface sites [38]. The pseudo-second-order model depends on the quantity of the adsorbent's surface. The second order equation is represented by Eqs. 3 and 4:

$$\frac{dQ_t}{dt} = k_2(Q_e - Q_t)^2 \tag{3}$$

$$\frac{t}{Q_t} = \frac{1}{Q_e}t + \frac{1}{k_2 Q_e^2} \tag{4}$$

k_2 signifies the pseudo second order constant.

3.3 *Adsorption Isotherms*

Not every adsorbent/adsorbate system behaves the same behavior. Adsorption phenomena are often approached by the isothermal behavior adsorbant/adsorbate. Isothermal curves describe the relationship at the adsorption equilibrium between the amount adsorbed and the solute concentration in a given solvent at a constant temperature.

The adsorption isotherm illustrates how a solute is distributed between the two phases, solid and liquid, when equilibrium is reached. Many theorems are used in order to explain the adsorption phenomena of a solute in a liquid or gas phase on a solid. Adsorption isotherms depend on temperature and concentration influence [26]. Among the various models developed in this context, we can report:

3.3.1 Langmuir Isotherm Model

The Langmuir model [48] indicates that the formation of a linear monolayer occurs when equilibrium is reached between the different solutes. All of the reactive sites on the surface have identical energy. The interaction forces reduce as the distance from the adsorption surface augment. The Langmuir equations are given by Eqs. 5 and 6:

$$Q_e = Q_m \cdot K \cdot (C_e/(1 + K \cdot C_e)) \tag{5}$$

$$\frac{C_e}{Q_e} = \frac{1}{Q_m} \cdot C_e + \frac{1}{Q_m \cdot b} \tag{6}$$

Q_e signifies the equilibrium capacity (mg/g), Q_m signifies the adsorption capacity of the adsorbent (mg/g), K signifies the Langmuir adsorption constant (L/mg) and C_e signifies the equilibrium concentration of the solute (mg/L).

3.3.2 Freundlich Isotherm Model

The Freundlich model [29] describes the relation between adsorbent amount and the solute concentration when equilibrium is reached. This model indicates non ideal and reversible adsorption. These are multilayers, with a non-uniform solute distribution sites for adsorption. The Freundlich equation is represented by Eqs. 7 and 8

$$Q = K_F \cdot C_e \exp(1/n) \tag{7}$$

Q_e signifies the equilibrium capacity (mg/g), K_F signifies the Freundlich constant, C_e signifies the equilibrium concentration (mg/L), and $1/n$ signifies the heterogeneity factor indicating the adsorption intensity.

3.3.3 Thermodynamic Adsorption Parameters

The transformation of a system is accompanied by a change in Gibbs energy ΔG. This variation depends on the initial and the final state. For a reaction of molecules to occur on a surface, the Gibbs energy is divided into two terms, an enthalpy ΔH which indicates the interaction energy between the solutes and the adsorbent surface, and an entropic term ΔS which indicates the modification and arrangement of solutes in the liquid phase. The viability of a reaction is defined by ΔG, which result due to the change in the enthalpy and the entropy [39]. A thermodynamic system always evolves spontaneously towards a lower energy level.

For a reaction to be possible in isolation, ΔG must negative. The Gibbs free energy (ΔG°), the change in entropy (ΔS°), and the change in enthalpy (ΔH°) are calculated [2, 68] using Eq. 9.

$$\Delta G = \Delta H - T\Delta S \tag{9}$$

4 Mesoporous Silica Materials

In the 90s, the mobil composition of matter, ordered mesoporous silica material, MCM-41 and the M41S family of mesoporous materials were discovered, the use of a mix of proprieties of surfactant with silicates, extended a new axis to exceed the pore size limit of zeolites.

Mesoporous silica materials have an inorganic nature, which are created by the reaction of sodium silicate with an ordered surfactant employed as a template. The synthesis of these mesoporous silica is influenced by some conditions, such as silicon nature, the morphology and nature of the surfactant, pH of the medium, temperature and time. These operating conditions directly affect the morphology and the composition of these mesoporous materials [18].

Mesoporous materials meet these parameters since they have a sizeable specific surface, narrow pore size distribution, stable arrangement and morphology, mechanical and chemical stability and plenty of Si–OH energetic linkages [3, 30, 31]. Porosity, which is the fraction of pore volume, is the major characteristic required for a solid to be used in catalysis, storage and separation, as the porosity gives rise to a high surface area of the material [32].

Mesoporous materials are categorized into three different types according to pore diameters of mesoporous materials. If the pore width is less than 2 nm the pores are called micro pores, if they are between 2 and 50 nm they are called mesopores, and if they are over 50 nm they are called macro pores. When different sized pores are mixed, we call the system a multi-scale porous network or Nano porous network.

The development of these porous materials inspired vast research on surfactant-directed assembly, employing surfactant to direct and assembly building blocks into mesoscopically ordered morphologies. Mostly, non-covalent interactions between the surfactant and the building blocks allow the building blocks to create supramolecular surfactant compositions first [40].

There have been increasing demands for new structures of porous materials after revealing that M41S materials have bad thermal properties. A novel material called santa barbara amorphous (SBA-15), was first obtained by the research group led by Stucky [72]. SBA-15 has attracted a considerable attention in the last ten years' due to excellent properties of material, such as: thermal stability, variable pore sizes, morphology and nature. The pore topology based of a two dimensional porous network of uniform dimensions composed by microporous walls.

Among all these solids, the best known is MCM-41 mesoporous silica which exhibits a hexagonal arrangement of regular pores [13]. Since this discovery, several other silicic materials santa barbara amorphous (SBA), korea advanced institute of science and technology (KIT), hexagonal mesoporous silica (HMS), were prepared according to the same mechanism by playing on the synthesis conditions and the compositions such as the type of silica precursor and the structuring agent.

MCM-41 is an ordered hexagonal mesoporous silicate. MCM-41 presents an important specific surface, high pore volume and uniform distribution of pore size. All these performances allow the use of MCM-41 in many applications such as elimination of heavy metal ions [4].

4.1 Synthesis Mechanisms

The preparation of mesoporous silica means the replication of a surfactant liquid crystal structure and the polymerization of a metal oxide precursor. Elimination of the surfactant through calcination leads to a porous structure supported by a hard silica framework [5]. Various researches have studied the formational process of surfactant template mesoporous silica. Researchers in mobil corporation proposed two models, namely: (i) the liquid crystal mechanism and (ii) the co-operative mechanism [14].

The synthesis of pure mesoporous materials strongly depends on the formation parameters such as temperature, time of reaction, pH, solvents, surfactant nature, hydrothermal treatment and calcination procedures [34]. Changing the synthesis pH leads to a change in the percentage of ionization of a surfactant, as exemplified by the phase transformation from higher curvature structures tetragonal or cubic to cylindrical hexagonal or cubic and to a lower curvature structure, lamellar [35].

Various synthesis mechanisms have been studied. This has obtained in series of mesoporous materials such as the M41S [14], folder sheet mesoporous (FSM-n) [40], hexagonal mesoporous silica (HMS-n) [62], SBA-n [60], fudan university, ordered mesoporous silica material (FDU-n) [69] and silica, mesostructured ordered mesoporous silica material (MSU-n) [15].

4.1.1 Structural Variations

A mesoporous silica material is defined as a material with a pore width between 2 and 50 nm, but a crystallographic definition of a mesoporous silica material is based on the arrangement of pores, ordered or disordered, in an amorphous framework. This arrangement of pores is complex, ranging from short interconnected one dimensional 1D pores, through 2D cylindrical pores to 3D interconnected pores [6]. Ordered porous materials possess narrow pore size distribution and have advantages in aspects such as tunable porosity, high surface areas, and broad range of molecular size for adsorption and for active compound release properties.

The characteristics of disordered porous materials, on the other hand, are governed by the randomness, connectivity and tortuosity of the pore space. Ordered mesoporous materials can be arranged according to structural dimensions and pore geometry of mesoporous materials.

4.2 Grafting Methods

Mesoporous materials have a structure with functional groups on the surface which facilitates the chemical modification of mesoporous materials. The functionalization of these materials' surface is accomplished by two ways; post-synthesis and co-condensation. Functionalization or chemical modification of mesoporous materials provides new structures and opportunity.

4.2.1 Post-Synthesis Method

Post-synthesis method consists of the modification of the surface of the mesoporous silica by grafting a functional group after the elimination of the surfactant [61]. Silanol groups existing on the surface and generally having a high concentration, thus being able to influence the process of functionalization of mesoporous materials. The process is based on the interaction of free groups and geminal silanol groups. However, silanol groups are much less reactive in the functionalization because they form hydrophilic networks between them [73].

4.2.2 Co-Condensation Methods

The functionalization from the co-condensation method is generally carried out by the sol–gel method using tetraalkoxysilanes and organoalkoxysilanes rich with Si–C bonds which allow the synthesis of hybrid inorganic–organic mesoporous materials [61].

The functionalization of mesoporous materials using the post-synthesis method involves grafting random functional groups onto the surface of the materials.

However, the co-condensation method ensures a homogeneous distribution of functional groups on the entire inner pore surfaces but no pore–blockage has been announced [70].

5 Applications of Ordered Mesoporous Materials

One fundamental property of a mesoporous silica material is the ability to provide space for functional organic groups on the silica wall. Mesoporous structures possess a high density of silanol groups after the elimination of a surfactant by calcination.

The reaction of the silanol groups with organoalkoxysilanes using the post-synthetic method leads to a random distribution (heterogeneous) of functional groups on the silica framework. However, by the simultaneous condensation of silica and organosilicas precursors in a single step using the co-condensation procedure a homogeneous distribution of functional groups on the silica wall may be achieved [71].

These processes provide several potentialities for choose these materials as drug carriers, adsorbents and catalytic supports [31, 32, 45]. If mesoporous materials are to be used as drug carriers, adsorptive properties such as pore width, surface area and pore volume are also the essential parameters to consider [64].

Adsorption and separation of gasses are another area attracting lot of attention. Surface functionalization of mesoporous materials with several types of functional groups for these applications has been recently reported.

The studies indicate that the co-condensation method leads to better results in terms of adsorption of gases or metal ions than post synthesis functionalization [64]. In addition to the quantity of surface functional groups, structural variations such as 2D cylindrical pores or 3D cage-type pores are also important. The 3D cage-type structures offer more advantageous characteristics than cylindrical pore structures such as the presence of cages and cage-connecting windows.

5.1 Applications of Functionalized Mesoporous Silicates in Removal of Heavy Metals

The affinity and selectivity towards targeted heavy metals in mesoporous materials can easy be improved by the grafting of adequate functional groups on the surface, which produces new adsorption sites [55]. A study performed by Lin and coworkers [49], shows that the surface of mesoporous materials was easily grafted with thiol groups and the adsorption capacity for the elimination of heavy metal was discussed.

Liu et al. [50] studied the synthesis of mesoporous grafted by amine and thiol groups. Authors was found that mesoporous which functionalized by amine groups had high levels of copper elimination [57] also indicated the high level of adsorption capacity of some heavy metals on silica Nano hollow spheres and silica gel after amine functionalization.

For example, the M41S family is formed of MCM-41, MCM-48, and MCM-50. However, MCM-41 is the category which has attracted researchers' attention the most and this is due to the properties such as; high pore volume, high surface area and mechanical and thermal stability [19, 20]. Also, MCM-41 is easily chemically modified by grafting functional groups on the surface which promotes the interaction between different metal pollutants and this mesoporous material for great elimination of ions from aqueous medium.

Following, some works from the last ten years concerning functionalized mesoporous materials which have been used as adsorbents for the removal of heavy metals from waste or contaminated water will be cited:

Costa et al. [21] were interested in the MCM-41 functionalized with aminobenzoic acid (PABA), otherwise known as PABA-MCM-41. Was employed for the elimination of the chrome in leather solution. PABA-MCM-41 was an electrostatic attraction between the positively charged chrome ions and the carboxylates groups of negatively charged PABA-MCM-41. The results showed that a removal percentage of 99% was found for the elimination of chrome ions from leather solution by the adsorbent synthesized.

The Fe_3O_4@MCM-41 nanocomposite was studied by Golshekan et al. [36]. The nanocomposite was charged with sulfonate groups by the obtaining of sulfonic functionalized organic–inorganic MCM-41 mesoporous material on the surface of nanocomposites. The introduction of zirconium oxide nanoparticles into the network of Fe_3O_4@MCM-41–SO_3H mesoporous silica nanoparticles elaborated by the covalently bound with implanted of charged sulfonate groups. The nanocomposite studied was employed for the preconcentration of lead from various samples of water.

Darzipour et al. [23] were interested in the SBA-15 functionalized with an antibacterial dynamic membranes immersed by cephalexin/amine. The mesoporous materials synthesized contributed to the elimination of heavy metal from aqueous medium. The mesoporous cephalexin/amine-SBA-15 nanocomposites was formed as a uniform and hydrophilic layer on the polyvinylidenefluoride ultrafiltration polymeric membrane. The synthesized material is efficient for the elimination of lead ions from samples of water.

Abedi et al. [7] reported the synthesis of monodispersed porous silica microspheres through the post grafting method and grafting of amine groups using triethylenetetramine. The new adsorbent synthesized is intended for cadmium ions removal from aqueous solution. During this study, the results showed that the grafting of amine groups decreased the pore volume, without affecting the pore diameter.

These results appear that the pore channel can be occupied by the amine functional groups. The deprotonating of the adsorption sites is promoted by grafting of the amine groups, consequently the pH of the medium increases. However, a high electrostatic attraction is created between the negatively charged adsorption sites and the positively charged metal ions which the adsorption process has promoted.

The co-polycondensation of alkoxysilanes using a sol–gel method was studied by [56] to obtain a nanocomposites functionalized with amino and mercapto functional groups. The magnetic nature of the nanocomposites ensures easy removal of materials from samples water. Using the post synthesis method, the porous structure was functionalized with 3-aminopropyltriethoxysilane and 3-mercaptopropyltriethoxysilane with different report. The various nanocomposites obtained have a specific surface area of 400 up to 800 (m^2/g). The synthesized composites mesoporous have a high sorption capacity to silver, copper, and lead ions.

The magnetic mesoporous silica nano-sorbent (MSN) was chemically modified with diethylenetriamine (DETA) to obtain a magnetic mesoporous material called (MSN-DETA). In fact, this study was carried out by [74]. Si–OH groups have a weakly acidic character which leads to a decrease in the interactions between these groups and the metal ions which are found in the aqueous solution, in order to improve this affinity Zhou and its collaborators have chosen the functionalization of the magnetic mesoporous silica with the amino groups of diethylenetriamine.

This synthesized material has given promising results in relation to the elimination of the heavy metals studied. The specific surface of the material synthesized was of the order of 153.68 mg/g and especially very high sorption capacity for uranium ions.

Sliesarenko et al. (2018) studied the use of sol–gel method to synthesize mesoporous silica modified with different concentrations of phosphonic acid groups on the surface to eliminate lead, cadmium, and dysprosium cations from aqueous solutions. The study showed that the addition of diethyl phosphatoethyltriethoxysilane influenced the pore size distribution, when the amount of diethyl phosphatoethyltriethoxysilane increases the pore size distribution also increases.

The phenomenon of adsorption was based on the formation of interaction bonds between the phosphonic acid groups grafted on mesoporous silica and metallic cations. The grafting of the phosphonic acid groups on the mesoporous solid has an effect on the character of the bonds formed.

El-Nahhal et al. [27] modified mesoporous SBA-15 silica materials with three compounds namely iminodiacetic acid, ethylenediaminetetraacetic acid and diethylenetriaminepen taacetic acid to get various grafted mesoporous. This involves grafting functional groups onto the SBA-15 in order to interpret the influence of the increase in amine groups in the grafting agents, such as a monoamine, a diamine and a triamine. The chemical modification of the adsorbent has proven to be very interesting by grafting several amine groups onto the surface especially for the elimination of the various metal ions studied from contaminated water.

An MCM-41 mesoporous material has been chemically modified using piperazine and the spinel cobalt ferrites by Kenani et al. [46]. The mesoporous material has been coated by the molecules of cobalt ferrites and grafted by the piperazine to increase the adsorption power of the functionalized material. Among the results found through this study, the adsorption capacities were 0.50, 0.30 and 0.25 μg/ L for lead, cadmium and copper, respectively. The magnetic power of the synthesized adsorbent can translate in the simple separation of the solid from the solution using a simple magnetic bar without the need for filtration or centrifugation, which saves time, expenses and effort.

Dindar et al. [24] were interested in the chemical modification of SBA-15 with the *n*-propylsalicylaldimine, the chemical modification was done using two grafting agents such as: aminopropyl to give SBA-aminopropyl, and ethylenediaminopropyl to give SBA-ethylenediaminopropyl. The adsorption process depends on the sensitivity of the medium to the pH value. The adsorption capacity of SBA-ethylenediaminopropyl has been valued for the elimination of metal ions.

The results have shown that the adsorption capacity increases up to a pH level of 3, while the adsorption capacity decreases. The adsorption of various cations on SBA-aminopropyl and SBA-ethylenediaminopropyl was also pH dependent. The phenomenon of interactions between adsorbent-adsrbat can be considered as an ion exchange. Through this study the best result was found by the SBA-aminopropyl for the elimination of lead ions from water.

Ezzeddine et al. [28], were interested in the functionalization of three mesoporous materials KIT-6, SBA-15 and SBA-16. The chemical modification was made with aminopropyltrimethoxy-silane and ethylenediaminetetraacetic acid using the post grafting methods. The synthetized materials were applied for the elimination of copper, nickel, cadmium and lead ions from samples of water. For KIT-6 and SBA-15, equilibrium was reached at about the first 20 min and the amount of copper ions removed was much higher than for SBA-16.

The ordered mesoporous silica KIT-6 was functionalized with carboxylic acid groups via the co-condensation method. The study was carried out by Bensacia et al. [16] and the synthesized adsorbent has been valorized for the elimination of lead cations from contaminated water. Through this study the grafting of the carboxylic groups was very promising.

In fact, the hydroxyl groups ionized in the solution forming showed negatively charged sites, this reinforces the negative charge on the surface by reacting with free and geminal silanol groups at the surface, which promotes interactions between the negative sites on the surface of the adsorbent and the positive charges of the lead cations. The functionalization of the kit-6 with carboxylic groups allowed an efficient adsorbent for lead ions.

In the same context, another study was carried out by Bensacia et al. [17], this was about a chemical modification for the mesoporous material called TUD-1, and this mesoporous material was grafted by the carboxylic groups using the post- grafting method. This functionalized mesoporous material was applied to remove cadmium cations from contaminated water.

Authors noted that there are three silanol groups on the surface of the mesoporous material. They are free, hydrogen-bonded hydroxyl and geminal groups. However, free and geminal groups are the only reactive groups. The number of these reactive silanol depended on the way the surfactant was eliminated and the operation condition was used. The carboxylic groups were obtained using hydrolysis of the nitrile groups using sulfuric acid.

A treatment of the pores of the mesoporous materials by some operations such as calcination, extraction and treatment with alcohol before the grafting reaction allows for the preservation of the free and geminal groups which plays a primary role in the adsorption process.

A study was carried out by [37] on the chemical modification of the mesoporous material SBA-15 by the grafting of amine groups, this mesoporous material grafted was applied for the removal of zinc, copper and cobalt ions from water. The SBA-15 was functionalized with 3-aminopropyl-triethoxysilane and trimethylpentyl phosphinic acid. Hydroxyl groups were reacted with the ethoxy groups of aminopropyl triethoxysilane and trimethylpentyl phosphinic acid.

Monitoring of the grafting reaction to the surface was carried out by the elemental and thermogravimetric analysis. The amino modified SBA-15 appeared high percentage of elimination for zinc ions compared to the other ions tested.

6 Conclusion

The development of the synthesis of mesoporous materials proved a very promising technology especially those environmentally friendly. Several studies have revealed the synthesis of new materials which are non-toxic, cost-effective and environmentally friendly to water pollution control. Mesoporous materials are blessed with a various physicochemical properties making them effective for the elimination of heavy metals.

Numerous studies on mesoporous materials as adsorbents reported that there can be promising solutions for heavy metal elimination from contaminated water. Since the mesoporous materials possess a large specific surface area, highly ordered pore arrangement and good mechanical and thermal stabilities. Also, the free surface silanol allow chemical modifications, which further increase the adsorbing power of these materials.

Mesoporous materials can be used as adsorbent for elimination toxic pollutants from wastewater with high efficiency able of being economically regenerated while maintaining a high adsorption capacity. The future of these materials in elimination of metal pollutants in wastewater is highly promising and has several opportunities.

References

1. Aroun I, Bensacia N, Seffah K, Benyahia S (2019) Kinetic and equilibrium studies of salicylic acid adsorption from contaminated water by (Alginate/Chitosan/Cobalt ferrite) Nanocomposites. Aljest 5(3): 1055–1061. www.aljest.webs.com
2. Atar N, Olgun A, Wang S (2012) Adsorption of cadmium(II) and zinc(II) on boron enrichment process waste in aqueous solutions. Batch and fixedbed system studies. Chem Eng J 192:1–7. https://doi.org/10.1016/j.cej.2012.03.067
3. Ardeshiri F, Akbari A, Peyravi M, Jahanshahi M (2019) Ahydrophilic–oleophobic chitosan/SiO_2 composite membrane to enhance oil fouling resistance in membrane distillation. Korean J Chem Eng 36:255–264. https://doi.org/10.1007/s11814-018-0188-4
4. Alejandra SD, Alvarez M, Volpe MA (2015) Metal-modified mesoporous silicate (MCM-41) material: preparation, characterization and applications as an adsorbent. J Braz Chem Soc 26(8):1542–1550. https://doi.org/10.5935/0103-5053.20150122

5. Atluri R, Bacsik Z, Hedin N, Garcia-Bennett AE (2010) Structural variations in mesoporous materials with cubic Pm3-*n*symmetry. Micropor Mesopor Mat 133(1–3):27–35. https://doi.org/10.1016/j.micromeso.2010.04.007
6. Atluri R (2010) Novel syntheses, structures and functions of mesoporous silica materials. Acta universitatis upsaliensis uppsala 2010. urn :nbn :se :uu :diva-122289ISSN
7. Abedi A, Taleghani HG, Ghorbani M, Kenari HS (2019) Facile and simple synthesis of triethylenetetramine—modified mesoporous silica adsorbent for removal of Cd(II). Korean J Chem Eng 36(1):37–47. https://doi.org/10.1007/s11814-018-0169-7
8. Boumalek W, Kettab A, Bensacia N, Bruzzoniti MC, Ben Othman D, Mandi L, Chabaca MN, Benziada S (2019) Specification of sewage sludge arising from a domestic Wastewater treatment plant for agricultural uses. Desalin Water Treat 143:178–183. https://doi.org/10.5004/dwt.2019.23559
9. Bensacia N, Moulay S (2012) Functionalization of polyacrylic acid with tetrahydroxybenzene via a homolytic pathway. Application to metallic adsorption. Int J Polym Mater 61(9): 699–722. https://doi.org/10.1080/00914037.2011.617343
10. Brião GV, Andrade JR, Silva MGC, Vieira MGA (2020) Removal of toxic metals from water using chitosan-based magnetic adsorbents. Environ Chem Lett 18:1145–1168. https://doi.org/10.1007/s10311-020-01003-y
11. Borji H, Ayoub GM, Bilbeisi R, Nassar N, Malaeb L (2020) How effective are nanomaterials for the removal of heavy metals from water and wastewater? Water Air Soil Pollut 231:330. https://doi.org/10.1007/s11270-020-04681-0
12. Bensacia N, Moulay S, Garin F, Fechete I, Boos A (2015) Effect of grafted hydroquinone on the acid-base properties of poly(acrylic acid) in the presence of copper(II). J Chem 2015:1–7. https://doi.org/10.1155/2015/913987
13. Behrens P, Stucky GD (1993) Ordered molecular arrays as templates: a new approach to the synthesis of mesoporous materials. Angew Chem Int Ed 32:696–699. https://doi.org/10.1002/anie.199306961
14. Beck JS, Vartuli JC, Roth WJ, Leonowicz ME, Kresge CT, Schmitt KD, Chu CTW, Olson DH, Sheppard EW, McCullen SB, Higgins JB, Schlenker JL (1992) A new family of mesoporous molecular sieves prepared with liquid Crystal templates. J Am Chem Soc 114(27):10834–10843. https://doi.org/10.1021/ja00053a020
15. Bagshaw SA, Prouzet E, Pinnavaia TJ (1995) Templating of mesoporous molecular sieves by nonionic polyethylene oxide surfactants. Science 269:1242–1244. https://doi.org/10.1126/science.269.5228.1242
16. Bensacia N, Fechete I, Moulay S, Hulea O, Boos A, Garin F (2014) Kinetic and equilibrium studies of lead (II) adsorption from aqueous media by KIT-6 mesoporous silica functionalized with–COOH. C R Chim 17(7–8):869–880. https://doi.org/10.1016/j.crci.2014.03.007
17. Bensacia N, Fechete I, Moulay S, Debbih-Boustila S, Boos A, Garin F (2014) Removal of cadmium(II) from aqueous media using COOH/TUD-1 Mesoporous solid. kinetic and thermodynamic studies. Environ Eng Manag J 13:2675–2686. http://omicron.ch.tuiasi.ro/EEMJ
18. Cecilia JA, Tost RM, Millán MR (2019) Mesoporous materials: from synthesis to applications. Int J Mol Sci 20:3213–3216. https://doi.org/10.3390/ijms20133213
19. Costa JAS, de Jesus RA, Silva CMP, Romão LPC (2017) Efficient adsorption of a mixture of polycyclic aromatic hydrocarbons (PAHs) by Si–MCM–41 mesoporous molecular sieve. Powder Technol 308:434–441. https://doi.org/10.1016/j.powtec.2016.12.035
20. Costa JAS, de Jesus RA, Santos DO, Mano JF, Romao LPC, Paranhos CM (2020) Recent progresses in the adsorption of organic, inorganic, and gas compounds by MCM-41-based mesoporous materials. Micropor Mesopor Mat. 291:109698. https://doi.org/10.1016/j.micromeso.2019.109698
21. Costa JAS, Costa VC, Pereira-Filho ER, Paranhos CM (2020) Removal of Cr (VI) from wastewater of the tannery industry by functionalized mesoporous material. Silicon 12:1895–1903. https://doi.org/10.1007/s12633-019-00315-1

22. Djillali Y, Chabaca MN, Benziada S, Bouanani H, Mandi L, Bruzzoniti MC, Boujelbene N, Kettab A (2020) Effect of treated wastewater on strawberry. Desalin Water Treat 181:338–345. https://doi.org/10.5004/dwt.2020.25095
23. Darzipour M, Jahanshahi M, Peyravi M, Khalili S (2019) Antibacterial dynamic membranes loaded by cephalexin/amine-functionalized SBA-15 for Pb(II) ions removal. Korean J Chem Eng 36(12):2035–2046. https://doi.org/10.1007/s11814-019-0391-y
24. Dindar MH, Yaftian MR, Pilehvari M, Rostamnia S (2015) SBA-15 mesoporous materials decorated with organic ligands: use as adsorbents for heavy metal ions. J Iran Chem Soc 12:561–572. https://doi.org/10.1007/s13738-014-0513-8
25. Engwa GA, Ferdinand PU, Nwalo FN, Unachukwu MN (2019) Mechanism and Health effects of heavy metal toxicity in humans. Intech Open, London. https://doi.org/10.5772/intechopen.82511
26. Ezzati R (2020) Derivation of Pseudo-First-Order, Pseudo-Second-Order and Modified Pseudo-First-Order rate equations from Langmuir and Freundlich Isotherms for adsorption. Chem Eng J 392:123705. https://doi.org/10.1016/j.cej.2019.123705
27. El-Nahhal IM, Chehimi M, Selmane M (2018) Synthesis and structural characterization of G-SBA-IDA, G-SBA-EDTA and G-SBA-DTPA modified mesoporous SBA-15 silica and their application for removal of toxic metal ions pollutants. Silicon 10:981–993. https://doi.org/10.1007/s12633-017-9556-7
28. Ezzeddine Z, Batonneau-Gener I, Pouilloux Y, Hamad H, Saad Z, Kazpard V (2015) Divalent heavy metals adsorption onto different types of EDTA-modified mesoporous materials: effectiveness and complexation rate. Micropor Mesopor Mater. 212:125–136. https://doi.org/10.1016/j.micromeso.2015.03.013
29. Freundlich HMF (1906) Over the adsorption in solution. J Phys Chem 57:385–470
30. Fechete I, Donnio B, Ersen O, Dintzer T, Djeddi A, Garin F (2011) Single crystals of mesoporous tungsten osilicate W-MCM-48 molecular sieves for the conversion of methylcyclopentane (MCP). Appl Surf Sci 257:2791–2800. https://doi.org/10.1016/S0169-4332(10)01838-6
31. Fechete I, Debbih-Boustila S, Merkache R, Hulea O, Lazar L, Lutic D, BalasanianI GF (2012) MnMCM-48, CoMCM-48 and CoMnMCM-48 mesoporous catalysts for the conversion of methylcyclopentane (MCP). Environ Eng Manag J. 11:1931–1943. https://doi.org/10.30638/eemj.2012.242
32. Fechete I, Ersen O, Garin F, Lazar L, Rach A (2013) Catalytic behavior of MnMCM-48 and WMnMCM-48 ordered mesoporous catalysts in a reductive environment: a study of the conversion of methylcyclopentane. Catal Sci Technol 3:444–453. https://doi.org/10.1039/C2CY20464A
33. Gupta VK, Srivastava SK, Mohan D, Sharma S (1998) Design parameters for fixed bed reactors of activated carbon developed from fertilizer waste for the removal of some heavy metal ions. Waste Manag 17:517–522. https://doi.org/10.1016/S0956053X(97)10062-9
34. Gross AF, Yang S, Navrotsky A, Tolbert SH (2003) In situ calorimetric study of the hexagonal-to-lamellar phase transformation in a nanostructured silica/surfactant composite. J Phys Chem B 107:2709–2718. https://doi.org/10.1021/jp0268563
35. Gao C, Qiu H, Zeng W, Sakamoto Y, Terasaki O, Sakamoto K, Chen Q, Che S (2006) Formation mechanism of anionic surfactant-templated mesoporous silica. Chem Mater 18:3904–3914. https://doi.org/10.1021/cm061107
36. Golshekan M, Shirini F (2020) Fe_3O_4@MCM-41-(SO_3^-) [ZrO_2] magnetic mesoporous nanocomposite: dispersive solid-liquid micro extraction of Pb^{2+} ions. Silicon 12:747–757. https://doi.org/10.1007/s12633-019-00143-3
37. Giraldoa L, Moreno-Piraján JC (2013) Study on the adsorption of heavy metal ions from aqueous solution on modified SBA-15. Mater Res 16(4):745–754. https://doi.org/10.1590/S1516-14392013005000051
38. Ho YS, McKay G (1999) Pseudo-second order model for sorption processes. Process Biochem 34:451–465

39. Imessaoudene D, Bensacia N, Chenoufi F (2020) Removal of cobalt(II) from aqueous solution by spent green tealeaves. J Radioanal Nucl Ch 324:1245–1253. https://doi.org/10.1007/s10967-020-07183-9
40. Inagaki S, Fukushima Y, Kuroda KJ (1993) Synthesis of highly ordered Mesoporous materials from a layered polysilicate. Chem Soc Chem Commun 8:680–682. https://doi.org/10.1039/C39930000680
41. Jadhav SA, Garud HB, Patil A, Patil GD, Patil CR, Dogale TD, Patil PS (2019) Recent advancements in silica nanoparticles based technologies for removal Of dyes from water. Colloid Interfac Sci Commun 30:100–118. https://doi.org/10.1016/j.colcom.2019.100181
42. kettab A (2017) Traitement des eaux usées urbaines et leurs réutilisations en Agriculture. Editions universitaires europeennes (2017–05–17)
43. Karbassian F (2018) Porous silicon. In: Ghrib TH (ed) Porosity-process technologies and applications. Intech Open, London. https://doi.org/10.5772/intechopen.72910
44. Kaonga CC, Kosamu IBM, Lakudzala DD, Mebewe R, Thole B, Monjerezi M, Crispine R, Chidya G, Keyili S, Mkali S, Sajudu I (2017) A review of heavy metals in soil and aquatic systems of urban and semi-urban areas in Malawi with comparisons to other selected countries. Afr J Environ Sci Technol 11(9):448–460. https://doi.org/10.5897/ajest2017.2367
45. Kim SN, Son WJ, Choi JS, Ahn WS (2008) CO_2 adsorption using amine-functionalized mesoporous silica prepared via anionic surfactant-mediated synthesis. Micropor Mesopor Mater. 115:497–503. https://doi.org/10.1016/j.micromeso.2008.02.025
46. Kanani, Bayat M, Shemirani F, Ghasemi JB, Bahrami Z, Badiei (2018) A Synthesis of magnetically N modified mesoporous nanoparticles and their application in simultaneous determination of Pb(II), Cd(II) and Cu(II). Res Chem Intermed 44:1689–1709. https://doi.org/10.1007/s11164-017-3192-0
47. Lagergren S (1898) For the theory of so-called adsorption of dissolved substances. K Sven Vetenskapsakad, Handl 24:1–39
48. Langmuir I (1916) The constitution and fundamental properties of solids and liquids Part I. Solids J Am Chem Soc 38:2221–2295. https://doi.org/10.1021/ja02268a002
49. Lin LC, Thirumavalavan M, Lee J F (2015) Acile synthesis of thiol-functionalized mesoporous silica- their role for heavy metal removal efficiency. Clean-Soil Air Water 43:775–785. https://doi.org/10.1002/clen.201400231
50. Liu A, Hidajat K, Kawi S, Zhao D (2000) A new class of hybrid mesoporous materials with functionalized organic monolayers for selective adsorption of heavy metal ions. Chem Commun 1:1145–1146. https://doi.org/10.1039/B002661L
51. Moulay S, Bensacia N (2016) Removal of heavy metals by Homolytically functionalized poly (acrylic acid) with hydroquinone. Int J Ind Chem 7:369–389. https://doi.org/10.1007/s40090-016-0097-5
52. Moulay S, Bensacia N, Garin F, Fechete I, Boos (2014) A synthesis of polyacrylamide-bound hydroquinone via a hemolytic pathway: application to the removal of heavy metals. C R Chim 17(7–8):849–859. https://doi.org/10.1016/j.crci.2014.03.011
53. Moulay S, Bensacia N, Fechete I, Garin F, Boos A (2013) Polyacrylamide—based sorbents for the removal of hazardous metals. Adsorpt Sci Technol 31(8):691–709. adt.sagepub.com/content/31/8/691
54. Masel RI (1996) Principles of adsorption and reaction on solid surfaces. Wiley, New York
55. Mubarak M, Alicia R, Abdullah E, Sahu J, Haslija AA, Tan J (2013) Statistical optimization and kinetic studies on removal of Zn^{2+} using functionalized carbon nanotubes and magnetic biochar. J Environ Chem Eng 1:486–495. https://doi.org/10.1016/j.jece.2013.06.011
56. Melnyk IV, Stolyarchuk NV, Tomina VV, Bespalko OV, Vaclavikova M (2019) Functionalization of the magnetite nanoparticles with polysilsesquioxane- Bearing N- and S- complexing groups to create solid-phase Adsorbents. Appl Nanosci 10:2813–2825. https://doi.org/10.1007/s13204-019-01087-1
57. Najafi M, Yousefi Y, Rafati A (2012) Synthesis, characterization and adsorption studies of several heavy metal ions on amino-functionalized silica nano hollow sphere and silica gel. Sep Purif Technol 85:193–205. https://doi.org/10.1016/j.seppur.2011.10.011

58. Pourret O, Hursthouse A (2019) It's time to replace the term *Heavy Matals* with *Potentially Toxic Elements* when reporting environmental research . Int J Environ Res Public Health 16(22):4446–4451. https://doi.org/10.3390/ijerph16224446
59. Seffah K, Bensacia N, Skender AE, Flahaut E, Hadj-ziane-zafour A (2017) Synthesis and characterization of nano-magnetic material based on (carbon nanotubes/nickel ferrite): application for the removal of methyl Orange dye from contaminated water. Aljest 3(1):45–53. www.aljest.webs.com
60. Sakamoto Y, Diaz I, Terasaki O, Zhao DY, Perez-Pariente J, Kim JM, Stucky GD (2002) Three-dimensional cubic mesoporous structures of SBA-12 and related materials by electron crystallography. J Phys Chem B 106:3118–3123. https://doi.org/10.1021/jp014094q
61. Stein A, Melde BJ, Schroden RC (2000) Hybrid inorganic–organic mesoporous silicates—nanoscopic reactors coming of age. Adv Mater 12:1403–1419. https://doi.org/10.1002/1521-4095(200010)12:19%3c1403::AID-ADMA1403%3e3.0.CO;2-X
62. Tanev PT, Pinnavaia TJ (1995) A neutral templating route to mesoporous molecular sieves. Science 267:865–867. https://doi.org/10.1126/science.267.5199.865
63. Verma R, Dwivedi P (2013) Heavy metal water pollution-a case study. Recent Res Sci Technol 5(5). https://updatepublishing.com/journal/index.php/rrst/article/view/1075
64. Vallet-Regi M, Balas F, Arcos D (2007) Mesoporous materials for drug delivery. Angew Chem Int Ed 46:7548–7558. https://doi.org/10.1002/anie.200604488
65. Walcarius A, Mercier L (2010) Mesoporous organosilica adsorbents: nanoengineered materials for removal of organic and inorganic pollutants. J Mater Chem 20:4478–4511. https://doi.org/10.1007/s12633-012-9122-2
66. WHO (2010) Ten chemicals of major public health concern. WHO, Geneva
67. Wu X, Cobbina SJ, Mao G, Xu H, Zhang Z, Yang L (2016) A review of toxicity and mechanisms of individual and mixtures of heavy metals in the environment. Environ Sci Pollut Res 23:8244–8259. https://doi.org/10.1007/s11356-016-6333-x
68. Young DM, Crowell AD (1962) Physical Adsorption of gases. Butterworth, London
69. Yu T, Zhang H, Yan XW, Chen ZX, Zou XD, Oleynikov P, Zhao DY (2006) Pore structures of ordered large cage-type mesoporous silica FDU-12s. J Phys Chem B 110:21467–21472. https://doi.org/10.1021/jp064534j
70. Yulin H (2009) Functionalization of mesoporous silica nanoparticles and their applications in organo- metallic and organometallic catalysis. Grad Theses Dissertations 10974. https://lib.dr.iastate.edu/etd/10974
71. Yokoi T, Yoshitake H, Yamada T, Kubota Y, Tatsumi T (2006) Amino—functionalized mesoporous silica synthesized by an anionic surfactant templating route. J Mater Chem 16:1125–1135. https://doi.org/10.1039/B516863E
72. Zhao XS, Ma Q, Lu VOCGQ (1998) Removal: comparison of MCM-41 with hydrophobic zeolites and activated carbon. Energ Fuel 12:1050–1053. https://doi.org/10.1021/ef980113s
73. Zhao D, Feng J, Huo Q, Melosh N, Frederickson GH, Chmelka BF, Stucky GD (1998) Triblock copolymer syntheses of mesoporous silica with Periodic 50 to 300 Angstrom pores. Science (Washington, DC) 279(5350):548–552. https://doi.org/10.1126/science.279.5350.548
74. Zhou L, Ouyang J, Liu Z, Huang G, Wang Y, Li Z, Adesina AA (2019) Highly efficient sorption of U(VI) from aqueous solution using amino/amine-functionalized magnetic mesoporous silica nanospheres. J Radioanal Nucl Chem 319:987–995. https://doi.org/10.1007/s10967-018-6381-4

Role of Water/Wastewater/Industrial Treatment Plants Sludge in Pollutant Removal

Shawani Shome, D. Venkatesan, and J. Aravind Kumar

Abstract Sewage sludges are obtained as a by-product from the process of the wastewater treatment plants. Production is anticipated to increase as asset in environmental organization rises and further municipal wastewater is treated to even advanced standards. The consistent development in the environmental impact of water resources is expected to be answered by a noticeable upsurge in sludge volumes formed. While there are numerous ways of placing sewage sludge, using it in water decontamination can turn it into a resource. Sadly, a large proportion of the sludge produced is not valorized but is excluded together with other residues in waste tips. The necessities of investigating probable innovative routes are obvious for sewage sludge valorization. Considering the account factors such as the existence of volatile components and the detail that sewage sludge is carbonaceous in nature, the sludge may be considered as potentially appropriate for the manufacture of activated carbon which are very useful in mixture separation and liquid purification due to their high adsorption capacity. Discarding of industrial wastewater poses a foremost environmental problem since such effluents comprise various pollutants that are resistant to conventional biological methods and thus difficult to remove. Re-use of that sludge may not only progress the particulate pollutant removal efficiency of a primary sewage treatment, but also ease the burden of water treatment works concerning to sludge treatment and clearance. This chapter aims to throw an insight on production rate, synthesis process, characterization and applications of sludge-based adsorbents (activated carbons).

Keywords Sludge based activated carbons (SBAC) · Sludge based adsorbents (SBA) · Brunauer Emmett Teller (BET) surface area · Wastewater treatment

---

S. Shome (✉) · D. Venkatesan · J. Aravind Kumar
Department of Chemical Engineering, Sathyabama Institute of Science and Technology, Chennai 119, India

E. Lichtfouse et al. (eds.), *Inorganic-Organic Composites for Water and Wastewater Treatment*, Environmental Footprints and Eco-design of Products and Processes,
https://doi.org/10.1007/978-981-16-5916-4_9

1 Introduction

The rising demands from environmental agencies and society towards better environmental quality standards have manifested themselves in private and public service administrators. As low indices of wastewater treatment exist in many developing countries, a future rise in the number of wastewater treatment plants is expected naturally. Therefore, the amount of sludge produced is also expected to rise. The municipal sewage sludge generation has also presently multiplied in congruence with rapid industrialization. Some green agencies in the developing countries now require the procedural definition of the final sludge disposal in the licensing processes. Hence, solid waste management is a burning matter of concern in many countries, tending towards a fast growing aggravation in the future years, as more wastewater treatment plants are implemented.

Wastewater treatment is the process of removing chemical, physical, biological contaminants and other pollutants and produce environmentally safe treated wastewater. The treated water can be further released back into nature. Sludge is a by-product of water, industrial and wastewater treatment operations. It is usually a semi-solid waste or slurry that has to undergo further treatment before being suitable for land application or disposal.

Sewage sludge is obtained from sewage treatment plants and consists of two basic forms, primary and secondary sludge. Biological treatment operations produce sludge which is also known as wastewater biosolids. The watery portion of the sludge is removed from liquid wastewaters containing less solid matter. The primary sludge includes precipitated solids that are generated during the initial treatment in the primary clarifiers. The secondary sludge parted in the secondary cleaners includes the sewage sludge purified from the secondary treatment bioreactors.

Sewage is generated by institutional, residential, industrial and commercial establishments. Above 99% of the sewage constitutes water that is a mixture of industrial and domestic wastes [1]. The difference between sludge and sewage is that sludge is a standard term for solids deducted from suspension in a liquid, while sewage is a suspension of solid waste and water, transported by sewers to be processed or disposed of. Conventional drinking water treatment and several other industrial processes produce sludge as a settled suspension.

Wastewater sludge includes a variety of inorganic and organic compounds. As per the recent researchers the use of sludge as an organic fertilizer in agricultural applications could ideally be a very attractive option [63]. For example, a large portion of insoluble aluminum hydroxides are contained by the aluminium-laden sludge which can be utilized as a coagulant in the primary sewage treatment [2]. However, there are some important drawbacks, which are mainly related to aspects such as appropriate soil availability, sludge quality and difficulties encountered in its monitoring and management.

2 Sludge and Its Production Rate in Wastewater Treatment Plants

The term 'sludge' refers to the solid by-products from wastewater/industrial/water treatment. Even though the sludge constitutes hardly 1–2% of the treated water volume, managing it is highly complex because it is frequently undertaken outside the boundaries of the treatment plant and usually costs 20–60% of the total operating costs of the wastewater treatment plant [3]. If the management of sludge generating from wastewater, industrial and water treatment plants is inadequately accomplished it may jeopardize the sanitary and environmental aspects in the treatment systems [4]. It is predicted that an average generation of around 50 g dry matter per inhabitant per day is archetypal for urban sewage plants and a corresponding rise in the quantity produced by industrial sewage plants. With high generation rate, together current and future estimated, the suitable management of sludges which are produced at sewage plants has become a need of the hour [5]. In the operational cost of wastewater treatment plants, the cost of waste sludge disposal is a major factor. Single handedly sludge dewatering constitutes around 40% of the annual operating costs. Although there are numerous ways of sewage sludge disposal, but rational practice of this left-over material can convert it into a resource.

2.1 *Stages of Sewage Treatment*

The Sewage treatment is mostly categorized into three phases: preliminary, primary and secondary treatment. Additionally, we can include two more stages of treatment for higher degree of separation depending on our needs [1].

2.1.1 Pretreatment or Preliminary Treatment

In preliminary treatment, girt and crude solids (diameter > 20 mm) are separated through screening. The obtained crude materials are not involved in biosolids. A sand or grit channel may be included in a pretreatment in which the velocity of the incoming sewage is adjusted such that it allows stones, sand, broken glass and girt to settle down. These crude solids are separated for the reason that their potential risk of damaging pumps and other equipment's. The feed in sewage water permits through a bar screen to subtract bulky objects like rags, cans, plastic packets and sticks carried in the sewage stream. These are collected and further incinerated in a landfill. Mesh screens or bar screens of variable sizes can be used for optimizing solid exclusion.

2.1.2 Primary Treatment

Suspended solids, grit and scum can be separated in the primary treatment by using following process of pre-aeration and sedimentation. The floating and settled substances are separated while the residual part may be released or exposed to secondary treatment. Oil and grease from the detached material can occasionally be separated for biodiesel production or saponification. With the help of air pumped via perforated tubes adjacent to the floor of the tanks, the wastewater is aerated making it less dense and causing the coarse solids settle out. As because the air jets are placed in such a way that the water is swirling while moving down the tanks, the suspended particles are inhibited from settling out. Dissolved oxygen is also provided by the air for the bacteria to use afterwards in the process. But for bacterial action to occur in the process, the wastewater in these tanks is not sufficient. The scrapers eliminate the solids from the tank bottom and the water jets washes off the scum. The solids and scum are brought to a common collection point wherein they are merged forming sludge and further forwarded to secondary treatment. In the primary sedimentation stage, sewage gushes through large tanks usually known as primary sedimentation tanks, pre-settling basins or primary clarifiers. Although the sludge settles down in the tanks, but the grease and oils mount to the surface and are skimmed off further. The primary wastewater treatment engages gravity sedimentation of the screened wastewater to separate the distorted solids. A fraction of the suspended waste stream passes through primary operation and discharges concentrated suspension as residue. This residue is also known as primary sludge and is further treated to yield biosolid.

2.1.3 Secondary Treatment

The secondary wastewater treatment is achieved through a biological process in which the biodegradable materials are removed. Microorganisms are used in this process to use up suspended and dissolved matter thereby producing carbondioxide and other byproducts. The density increases when the microorganisms are added. The cleaned water is then separated resulting to a formation of a concentrated suspension called secondary sludge. This whole process takes place at the bottom part of the water tank. It is necessary to separate the microorganisms from the water before releasing it or sending it for tertiary treatment. In secondary treatment the organic content of the sewage which is derived from various sources are considerably reduced. The nutrients required to uphold the microorganism population is supplied through the organic material and the ones within the sludge are converted to carboxylic acids and further to carbondioxide by aerobic fermentation or methane by anaerobic fermentation. The biogas thus obtained is an important source of fuel. The volume of the sludge leaving the digesters reduces up to its half. Mostly the municipal plants treat the settled sewage water through aerobic organic processes. The biota requires both food and oxygen for living efficiently. Inclusion of secondary clarifiers in secondary treatment processes helps in settling down the organic flocculated substances developed inside the bioreactor. Many designs of hybrid treatment plants are produced to treat tough wastes, consume less space and for intermittent flows also.

2.1.4 Tertiary Treatment

Further processing is necessary in case of high quality waste requirement such as discharging them directly to the drinking water source. The tertiary treatment usually yields a solid residue which primarily includes the chemicals added to the raw waste before evacuating and thus it is not regarded as a biosolid. If the whole treatment process is minutely managed throughout then plant operators can control nutrients, solid ingredients and various other components of biosolids. For municipal biosolid production, majority of the material used is through operating primary and secondary effectors simultaneously. The prime reason of tertiary treatment is to improve the effluent quality before discharging it to the environment. There may be requirement of more than one tertiary treatment processes. No matter what, disinfection is always considered the last process and is usually known as effluent polishing.

2.1.5 Fourth Treatment Stag

Particles of chemicals used in small industries, households, pharmaceuticals or pesticides are considered as micro-pollutants and they may not be able to be separated through the usual treatment process like primary, secondary and tertiary treatment. Also if they are not separated it may lead to water pollution. Therefore, a separate treatment named fourth treatment stage is introduced in the sewage treatment process to remove the micro-pollutants. These techniques are not yet functional on a regular basis as they are still very expensive.

2.2 Volume Reduction Processes

The quantity of wastewater treatment plant sludge produced can be expressed in terms of volume (wet basis) or mass (dry basis). The sludge production can be expressed in an easy way in terms of per capita and chemical oxygen demand (COD) bases for mass and volume calculations. The organic sludge is generally made up from the biomass which is formed from the conversion of some of the COD in the biological wastewater treatment. This sludge is actually produced from secondary sludge and thus has suspended solid composition is less than 1% (wt.). However, primary sludges are high concentrated and the combination of both primary and secondary sludge of solid concentrations contains around 3% by weight. As the sludges are naturally in voluminous, treatment processes are named as dewatering, thickening, conditioning and drying. Water removal helps in improving the efficiency of the further treatment processes, reducing the storage and decreasing transportation costs.

2.2.1 Thickening

A concentrated product is produced in the sludge thickening process and that product basically retains the liquid properties. The most common thickening process usually applied to municipal sludges is concentration by simple sedimentation or gravity thickening. The sludge product obtained from gravity thickening. Practice of centrifuges, gravity drainage belts, perforated rotating drums and floatation method are alternative to gravity thickening. The product from gravity sludge thickening frequently comprises around 5–6% solid by weight. Floatation is a process wherein a gas is included in sludge solids which results in making them to float.

2.2.2 Dewatering

In the sludge dewatering process, the product ensures solid properties even though the water content retained in it is still not negligible. The thickened sludge is transported through tank truck but in case of dewatered sludge a dump truck is used. Sand drying beds and at times lagoons are used for dewatering process but removal of moister in thickening step is enabled through gravity drainage and sedimentation. Mostly, dewatering mechanical sludge equipments namely, vacuum filters, belt filter press, centrifuges and filter press are used in big municipal installations. Mechanical method is more efficient than other processes because the solid content of the product sludge by weight ranges from 20–45% which is quite higher.

2.2.3 Conditioning Sludge

Conditioning process does not reduce the moisture content directly, rather changes the physical and chemical properties of the sludge which helps it in water discharging in dewatering process. Without the prior conditioning of the sludge, mechanical dewatering process will not be economical. In chemical conditioning, mostly synthetic organic polymers or inorganic chemicals like ferric chloride and lime are added to the sludge before dewatering. The mass of the solid sludge increases because of the huge dosages of inorganic chemical conditioning. Physical conditioning on the other hand includes freeze–thaw treatment and heat treatment.

2.2.4 Drying

Drying step is usually required if the need for further water removal arises even after dewatering step. Thermal drying in association with indirect or direct driers are generally used to attain almost entire water exclusion from sludges. Also, solar drying can be an option in several locations. The heat generated in biochemical reactions during composting and various other chemical reactions results in partial drying.

2.3 *Stabilization Processes*

Sludge stabilization is practiced to reduce the problems aroused due to biodegradation of organic substances and is generally performed through chemical and biological treatment methods. The vector attraction reduction provision of the Part 503 Sludge Rule [EPA, 1993a] is concerned in Stabilization processes. Vectors are nothing but organisms that may get fascinated to sludges which are not stabilized and may lead to spread of infectious diseases. While applying sewage sludge to the agricultural lands if we inject it below the surface or into the soil then the vector attraction can be minimized. The sludges can also be stabilized by drying it adequately to obstruct microbial action. Combusting the sludge can also facilitate its stabilization. Various stabilization processes can also inactivate pathogenic organisms and viruses.

2.3.1 Biological Stabilization

In this process, the biological sludges are cut down through biological degradation processes in a exact and well-engineered manner. Methane is yielded as a byproduct when the household wastewater sludge is stabilized biologically in the form of a liquid inside anaerobic digesters. This liquid sludge also can be stabilized biologically in an aerobic digester in presence of oxygen. Composting is an aerobic process which facilitates in biological stabilization of the dewatered sludge. This process takes place in thermophilic temperature of around 55 °C due to heat released during biochemical transformations. Sawdust and Wood chips should be additional to progress friability to encourage aeration. The heat from the same source can be used in case of operating the aerobic digesters thermophilically.

2.3.2 Chemical Stabilization

In this process the sludges not only intend to reduce the biodegradable organic matter quantity and also generating suitable conditions for the inhibition of microbial activity in order to prevent odors. Out of all the chemical stabilization methods available the most common one is raising the pH value of the sludge using cement kiln dust and lime. Liquid or dewatered forms of sludge can be chemically stabilized. During the chemical stabilization of dewatered sludge an exothermic reaction between lime and water results in heating that facilitates in pathogen destruction and water evaporation as well.

3 Preparation of Adsorbents Derived from Sludge

Lately liquid-phase adsorption is popping up as a promising option to remove the non-biodegradable pollutants from water streams. The most common adsorbent for the liquid phase adsorption is activated carbons because of their versatility and effectiveness. Small particle sized adsorbents are preferred in case of solution phase because of their large surface area and results in small diffusion distance. The activated carbon prepared by treating wastewater sludge is a black amorphous can be used to treat pollutants. The wastewater sludges are blended, carbonized, activated and acid or alkaline treated to form activated carbon. The safe eco-friendly sludge based activated carbon has higher absorbability due to its dense pore characteristic with more specific surface area and complex structure. It has wide variety of raw material source, good thermal stability, even chemical properties and can also be recovered and utilized repeatedly.

The activated carbons can be prepared from sludge through various methods like physical activation, chemical activation, physical–chemical activation, direct pyrolysis, microwave activation etc. All these methods are used to produce porous carbonaceous adsorbents.

The alteration and synthesis pattern of sludge-based activated carbon is concisely explained below. To obtain sludge based activated carbon, the activated sludge can be dried and pulverized directly under inert gas. This is called direct pyrolysis and it takes place in the following stages:

(i) The drying is the first stage
(ii) The second stage is the pyrolytic wherein huge amounts of volatile components are melted.
(iii) The last stage is where the remaining material carries on pyrolysing gradually.

Through physical activation, the ground activated sludge can be directly pyrolysed and dried under inert gas protection and further pyrolysed to finally obtain the sludge based activated carbon under various other protective gases like water vapor, carbondioxide and flue gas. The traditional Muffle furnace heating method is often used in physical activation of the sludges.

In chemical activation method, the raw sludge components are either put together with chemical reagents at a ratio or the sludge is dip dried in a solution of chemical reagent as per a definite solid-liquid ratio and further the mixture is pyrolysed to get our desired product.

Physical–chemical activation is the combination of both physical and chemical activation wherein the sludge is mixed with chemical reagent in a certain ratio and further they are pyrolysed under the protection inert gas to yield sludge based activated carbon.

In microwave activation the sludge is pyrolysed and carbonized into activated carbon through microwave heating. Microwave activation has recently drawn consideration since it is easy to control, energy consumption is low, high efficiency, cost effective, less pollution and more feasibility. As the carbon content in these activated carbons are low therefore carbon source materials like wood chip, corn kernel, peanut and hazelnut shells are added to increase the carbon source so that it can exhibit abundant pore structure and high adsorption efficiency [3].

4 Sludge-Based Adsorbents Characterization

The efficiency of sludge based adsorbents in removing the contaminants is decided by their surface and structure chemistry features. The Brunauer Emmett Teller (BET) surface area is the most common analysis to assess the structure of an adsorbent. The Barrette Joynere Halenda (BJH) method helps in calculating the pore size distribution, macropore and mesopore volumes. The t-plot method on the other hand helps in calculating the micropore volume. In the following sections the effects of pyrolysis condition on the structure of the sludge based adsorbents are discussed [4].

4.1 Carbonization

The Table 1 tabulates the pore structures and BET surface areas of sludge-based adsorbents when they undergo only carbonization. The various influences in carbonization that affects the chemical and physical features of sludge-based adsorbents are pyrolysis temperature, dwell time, feedstock type and heating rate. These are discussed below

4.1.1 Pyrolysis Temperature

The temperature of pyrolysis plays an important role in altering the characteristics of sewage sludge and industrial sludge adsorbents. As per our observations, an increase in pyrolysis temperature rises the ash content of these adsorbents but decreases its yield too [6, 7]. At high pyrolysis temperature, devolatilization of the solid hydrocarbons and the integrant gasification of the carbonaceous residues in the adsorbents take place [8]. Pyrolysis temperature also affects the surface acidity or basicity, morphology and surface characteristics changes in sludge based adsorbents. Usually, the adsorbents that are produced at high temperature of around 500 °C are alkaline and those in low temperature are acidic in nature [6]. At high temperature sodium oxide is released from the sludge which increases its alkalinity [9].

Generally, with increase in pyrolysis temperature the pore volume surface area and the BET surface area also increases. But when the temperature is excessively

Table 1 Pore structure characteristics of sludge based adsorbents produced by only carbonization [4]

Type of Sludge	CARBONIZATION			POST TREATMENT	BET Surface area (m^2/g)	Total pore Volume (cm^3/g)	Micropore Volume (cm^3/g)
	Dwell Time (hr)	Temperature (°C)	Heating rate (°C/min)				
Wastewater treatment plant	1.5	450	5	Hydrochloric acid	15	0.02	–
Wastewater treatment plant	0.5	650	40	–	60	0.04	0.05
Electroplating sludge	1	500	10	Water	19.6	-	–
Electroplating sludge	1	950	10	–	127	0.158	0.054
Paper mill sludge	2	650	3	–	275	0.017	0.011
Sewage and waste oil sludge	0.5	650	10	–	108	0.043	0.313
Sludge and disposable filter cake	1.5	450	5	Hydrochloric acid	60	0.1	-

high, destruction of porous structure takes place and its combination of mesopore inhibits further development of porosity [10]. This increase in surface structure on high pyrolysis temperature is resulted due to the increase in the degree of atomization and the rearrangement in nitrogen chemistry [11]. The porosity is increased through carbonization due to the mass loss during the thermal decomposition and evolution of volatile matter [6]. Additionally, the creation of micropore is also boosted when the high moisture content of wet sludge generates a steamy atmosphere at high temperature resulting to partial gasification of the solid char [8]. On the other hand, very high temperature can probably lead to a decrease in surface area because of the porous structure destruction development of deformation, cracks or blockages of micropores in those adsorbents [10]. The optimum carbonization temperatures for increasing the BET surface areas to maximum are reported as 450, 500, 550, 650 and 950 °C.

4.1.2 Dwell Time

Dwell time is the period of time that an element or system remains in a given state. According to the studies, higher pyrolysis temperature results in shorter dwell time. The optimum dwell time estimated at 500 °C, 650 °C and 950 °C are 240,

120 and 60 min respectively [9, 11–13]. Also it has been found that at 650 °C with increase of dwell time the micropore volume of sludge based adsorbents remained constant while the surface area and mesopore volume decreased [11]. Some experiments have also deduced that at the same pyrolysis temperature of 650 °C, with increase of dwell time the micropore volume decreases [14].

4.1.3 Heating Rate

According to the studies, higher heating rates improves the product yield and carbon content but decreases the hydrogen content of the sludge based adsorbents. Whereas, lower heating rates of about 3 °C/min increases the BET surface area [9, 15]. It was probably due to larger sample residence time during the pyrolysis processes. As per Table 1, the heating rates are 3, 5, 10, 20 and 40 °C/min.

4.1.4 Type of Feedstock

Researches implies that addition of high carbon content materials like leaf litter, disposal filter cake, waste oil sludge and solid residue of pyrolysed tyres to the sludge can improve their porosities [13, 16, 17]. Due to the oil volatilization and hydroxide formation during pyrolysis, the addition of waste oil sludge in sewage sludge in the mass ratio 1:1 improves the BET surface areas and micropores volume [11]. Also the adsorbent yielded by mixing the sewage sludge and filter cake in the mass ratio 85:15 has higher BET surface area of about 60 m^2/g than the regular sludge based adsorbents which has a BET surface area of about 15 m^2/g [16].

4.2 Physical Activation

The process of physical activation generally takes place in the following two steps:

The first step is to carbonize the sludges around the temperature between 400–700 °C in presence of inert gas like nitrogen or helium to break down the cross-linkage bonds among carbon atoms. The next step is to activate it with the help of gases like nitrogen, oxygen or air, steam, carbondioxide etc. at a high temperature of about 800–1200 °C. This facilitates in developing the porosity of the sludge based adsorbents further. The most common activator gases are steam and carbondioxide. In the Table 2 a summary of the characteristics of sludge based adsorbents that are activated at various conditions.

Table 2 Pore structure characteristics of sludge based adsorbents produced by carbonization and physical activation [4]

Type Of sludge	CARBONIZATION			PHYSICAL ACTIVATION					**Post Treatment**	BET Surface Area (ml/min)
	Dwell Time (hr)	Temperature (°C)	Heating rate (°C/min)	Activator Gas	Flow rate (ml/min)	Dwell Time (hr)	Temperature (°C)	Heating rate (°C/min)		
Cosmetics sludge	Not carbonized	800	100	Carbondioxide	10	2	100	94	Hydrochloric acid	0.04
Cosmetics sludge	Not carbonized	750	100	Nitrogen	10	0.5	100	44	Water	0.063
Paper mill sludge	Not Carbonized	650	275	Oxygen/air	3	2	–	–	–	0.061
Paper mill sludge	Not Carbonized	600	300	Carbondioxide	10	-	300	17	–	–
Methane fermentation sludge	1	500	244.6	Nitrogen	10	1	700	10	Water	10
Wastewater treatment plant	0.5	950	269.1	Steam	–	1.6	900	10	–	10
Wastewater treatment plant	Not carbonized	750	-	Nitrogen	20	–	0.5	34.3	–	-

4.2.1 Steam

As compared to the activation using carbondioxide, the steam activation at a given temperature, leads to larger adsorption capacity and wiser pore size distribution in the sludge based adsorbents. This method encourages the micropore and mesopore creation [18]. Steam activation mechanism is the collective effect of fixed carbon loss and devolatilization that results from water gas reaction in this case. With increase in activation temperature, the BET surface areas of adsorbents prepared by steam activation also rises up due to the higher rate of diffusion of the water molecule to the inside thereby broadening the pore network [19, 20]. But if the activation temperature crosses 850 °C, BET surface areas decreases as more particles start burning out.

4.2.2 Carbondioxide

The porosity of the sludges can be enhanced with the help of carbondioxide activation. The development of opened micropore and closed opening of micropore takes place by removing the carbon atoms from the interior of the particle through gasification at higher temperatures of about 900–1200 °C and lengthier dwell time [21]. During physical activation, the development of porosity of sludge based adsorbents is restricted if the raw material contains high ash content. It is observed that the acidity of the sludge-based adsorbents prepared through carbondioxide activation declines with an inclination in the activation temperature because of the oxygenated acidic surface group's degradation [22].

4.3 Chemical Activation

Chemical activation is nothing but activating the sludge based adsorbents by chemical treatment at specified conditions. The factors that affect the chemical activation are activator kinds, activator temperature, activation concentration, addition of binder. The Table 3 describes the pore structure characteristics of sludge based adsorbents by chemical activation.

4.3.1 Types of Activator and Activation Temperature

Activator plays the most important role in influencing the processes of chemical activation [23]. Various activators like sulfuric acid, phosphoric acid, potassium hydroxide, sodium hydroxide, zinc chloride, ferric chloride and potassium carbonate can be used but zinc chloride, sodium hydroxide, potassium hydroxide and phosphoric acid are the most common used ones.

According to the observations tabulated in Table 3, potassium hydroxide has proved to be the most effective activator as it has produced sludge based adsorbents

Table 3 Pore structure characteristics of sludge based adsorbents produced by carbonization and chemical activation [4]

Type of sludge	CARBONIZATION			CHEMICAL ACTIVATION					POST TREATMENT	BET Surface Area (m^2/g)	Total Pore Volume (cm^3/g)	Micropore Volume (cm^3/g)
	Dwell time (hr)	Temperature (°C)	Heating rate (°C/min)	Activator gas	Mass ratio	Dwell time (hr)	Temperature (°C)	Heating rate (°C/min)				
Cosmetics sludge	Not carbonized			Potassium hydroxide	1:1	0.5	750	10	Hydrochloric acid	950	0.4	0.23
Waste water Treatment plant	Not carbonized			Potassium hydroxide	1:1	1.0	500	-	water	69	–	–
Waste water treatment plant	0.5	700	5	Potassium hydroxide	1:1	1.0	700	5	Hydrochloric acid	1882	0.89	0.67
Waste water treatment plant	Not carbonized			Sodium hydroxide	1:25 M	1.5	850	10	Hydrochloric acid	422	NA	–
Waste water treatment plant	Not carbonized			Sodium hydroxide	1:1	3	700	–	water	139	0.06	–
Waste water treatment plant	1	500	5	Sodium hydroxide	1:1	2	600	5	Sodium hydroxide	121	0.10	–

(continued)

Table 3 (continued)

Type of sludge	CARBONIZATION			CHEMICAL ACTIVATION					POST TREATMENT	BET Surface Area (m^2/g)	Total Pore Volume (cm^3/g)	Micropore Volume (cm^3/g)
	Dwell time (hr)	Temperature (°C)	Heating rate (°C/min)	Activator gas	Mass ratio	Dwell time (hr)	Temperature (°C)	Heating rate (°C/min)				
Waste water treatment plant	Not carbonized			Phosphoric acid	3 M	0.167	800	–	Sodium hydroxide	291	–	–
Waste oil sludge	Not carbonized			Zinc chloride	1 M	0.5	500	15	water	737	0.21	–
Waste water treatment plant	Not carbonized			Zinc chloride	5 to 30%	1.5	375	–	Hydrochloric acid	270	0.05	–
Waste water treatment plant	1.0	450	-	Zinc chloride	6 M	2.0	750	10	Hydrochloric acid	510	0.16	–
Waste water treatment plant	Not carbonized			Zinc chloride	1:1	0.5	650	40	Hydrochloric acid	472	0.04	0.05

(continued)

Table 3 (continued)

Type of sludge	CARBONIZATION			CHEMICAL ACTIVATION					POST TREATMENT	BET Surface Area (m^2/g)	Total Pore Volume (cm^3/g)	Micropore Volume (cm^3/g)
	Dwell time (hr)	Temperature (°C)	Heating rate (°C/min)	Activator gas	Mass ratio	Dwell time (hr)	Temperature (°C)	Heating rate (°C/min)				
Waste water treatment plant	Not carbonized			98% Sulfuric acid	1:1	0.5	650	40	Hydrochloric acid	–	–	-
Waste water treatment plant	Not carbonized			Potassium carbonate	1:1	1.0	800	20	water	422	–	–
Waste water treatment plant	1.5	500	20	Nitric acid	7 M	2	750	15	–	273	0.19	0.06
Methane fermentation sludge	Not carbonized			Ferric nitrate	2:1	1	700	10	–	245	0.24	0.12

with high BET surface area with a value as high as 1882 m^2/g [24]. This process is high energy consuming and the product can be obtained through a two-stage method—carbonizing before impregnating and activating it while activator to solid ratio is maintained as 1:1. Also as per the research conducted, sludge based adsorbent prepared through single stage method with potassium hydroxide maintaining activator to solid ratio as 3:1 yields a value of BET surface areas as 1832 m^2/g [25]. The mechanism of potassium hydroxide activation is that an intercalation compound of carbon and potassium oxide is formed which infiltrates inside and at high temperature this potassium oxide reduces to metallic potassium atoms [26]. This results in gasification and emission of steam and carbondioxide which facilitates in pore formation. Also, potassium vapor widens the gap between carbonaceous layers and thus increases the surface area.

Zinc chloride is another such effective activator with yields BET surface area as high as 757m^2/g [27]. It helps in dehydrating and forming tar to suppress the activation process and also promotes carbon skeleton aromatization to form pores [28]. As per Table 3, the optimum activation temperatures for zinc chloride depending on the feedstock types are reported as 300, 375, 500 and 750 °C. A washing step can create extra micro and mesoporosity to remove the zinc chloride and zinc oxide entrapped.

Although as per Table 3 the BET surface area of sludge based adsorbents activated through phosphoric acid is merely 290.6 m^2/g but the only advantage it has its low cost and activation temperature. During activation processes the effects of phosphoric acid are dehydration, depolymerization, rearrangement of biopolymers constituent, specifically, encouraging the change of aliphatic to aromatic compounds thereby growing the yield of solid phase products.

4.3.2 Concentration of Activator

The optimum value of activator concentration be subject to on its characteristics and the type of feedstock. Usually with the increase of activator concentration, the BET surface areas and adsorption capacities increase but if it exceeds appropriate values the BET surface areas and adsorption capacities start decreasing due to partial destruction of microporosity resulted from hyper-activation [29].The optimum activation concentration for potassium hydroxide, sodium hydroxide and zinc chloride as per research are 1 M, 1.25 M, 2 M respectively [29, 10].

4.3.3 Binder Addition

Although adding binders to the sludge based adsorbents prior to chemical activation produces huge granules but reduces the surface areas of BET [30]. In catalytic wet air oxidation process these hard adsorbents can be used to increase the pollution removal rate [31].Phenolic resin, clay, polyvinyl acetate (PVA), lignosulphonate, humic acid are the most commonly used binders. Clay, phenolic resins and humic acid decreased the micro and macroporosity of sludge based adsorbents [32]. Although humic acid and phenolic resin addition had negligible effect on BET surface areas but in case of clay it decreased to a considerable extent. While sludge based adsorbents formed without binder are friable with a hardness number in the range of 58–71% those produced in a combination of steam activation and polyvinyl acetate binder (5 wt %) yields hardness number of around 92–93% [33].

4.4 Post Treatment

Post treatment, implies to treatments such as acid washing, alkaline washing or distilled water washing which helps in decreasing the ash contents, increasing the BET surface areas and porosity and even removing the extra reaction products and activation agents in case of chemical activation. An appropriate ash-dissolution technique like washing with hydrochloric acid can be applied to reduce this high ash content [33]. Acid Washing is the most widely used one among them as it cut downs the inorganic content of the carbonaceous material by dissolving the basic oxides like aluminium oxide, ferric oxide, calcium oxide etc. from the adsorbent to increase porosity [34].

5 Applications

The sludge based adsorbent has found its demand in various sectors these days. They are explained as follows [3]:

5.1 Organic Matter Removal

In the Table 3 it is shown that the organic matters like phenol, toluene, trinitrotoluene, nitrobenzene, rhodamine B and ibuprofen can be removed with the sludge based adsorbents over physico-chemical adsorption and hydroxyl radical oxidation [5, 36–40]. The sludge based adsorbents prepared through phosphoric acid microwave method was used to remove trinitrotoluene from water and from the previous research it was concluded that it has pretty bulky surface and plentiful extension holes [38]. When more amount of aluminum oxide and iron is added to the sludge based adsorbents then the deletion rate of UV254 and dissolved organic carbon by it are around 85.8% and 59.7% respectively which is almost similar to the commercial ones [41]. When suitable amount of raw materials like sawdust, corn cobs, coconut shell activated by zinc chloride is added to dehydrated sludge the adsorption efficiencies generally increases [42, 43]. In the toluene adsorption experiment it was shown that the equilibrium efficiency of shell sludge based adsorbent is the highest followed by coal adsorbent and sawdust adsorbent. Similarly, for phenol and nitrobenzene removal, sludge based adsorbents with corn cores were used which proved that higher doping proportion of corn cores results in larger micropore volume, BET surface area [44].

These adsorbents prepared from sludge can also act as catalyst or carrier to prepare necessary conditions for new composite photocatalytic material preparation [45, 46]. As per the studies, the sludge based activated carbons mixed with oxides of manganese has decent catalytic activity [47]. The reaction mechanisms like hydroxyl radical reactions and surface reaction were involved in the catalytic ozonation process of oxalic acid mineralization. For increasing the productivity of ozone oxidation of wastewater pollutants, transition metals like iron oxide and manganese were usually doped into the sludge based adsorbents through impregnation method [48]. Due to various modification methods involved, the principles for organic matter removal are at times different from one another. In the catalytic ozone oxidation of rhodamine, the sludge based activated carbon prepared with a mixture of biological and chemical sludge follows the mechanism of hydroxyl radical oxidation [49].

Sludge based adsorbents can also be mixed with Fe_3O_4 or metal free materials like nitrogen rich urea and these are named as F-SBAC and N-SBAC respectively [50, 51]. The F-SBACs produced hydroxyl radicals to catalyze hydrogen peroxide and it can be prepared at different temperatures like 600, 800 and 1000 °C and based on it the surface area, porous structure and removal rates differed. The chemical microenvironment and microstructure is influenced by N-SBAC and to remove the organic contaminant it can oxidized effectively too. For adsorbing phenol, the sludge based activated carbon followed electron donor receptor reaction mechanism among the aromatic phenolic rings and the adsorbent surface functional groups [52]. It is estimated that either due to the competition between the two composites at the surface adsorption sites of the sludge based activated carbons or due to the space resistance of the co-adsorbent phenol, its adsorption capacity in removing cadmium ions is decreased.

As per the research works, citric acid-zinc chloride mixed with sludge based adsorbent seemed to be a good green technique in char manufacture with good pore structure [53]. The sludge derived char is a hybrid material that contains carbon in elemental form, aromatic organic matter and inorganic ash and this char can also treat various kinds of benzene derivatives in aqueous solution.

Dibenzothiophene (DBT) can also be removed from n-octane by these sludge based adsorbents and its adsorption rate increases with increase oxygen-containing functional groups like carbonyl groups [54]. Out of the activator used in DBT removal, potassium hydroxide enabled the highest adsorption capacity even more than commercial activated carbons.

5.2 Heavy Metal Removal

The sludge based activated carbons helps in removing heavy metal ions by surface precipitation, ion exchange reaction, chemical and physical adsorption [55]. These metal ions tend to create an exchange reaction on the surface of activated carbon. After modifying or adding a reagent to the sludge based activated carbons, its surface can get exposed to special surface groups to strengthen the absorption of heavy metal ions and form products. These special groups develop into ligands with heavy metal ions. The type and stability of these ligands helps in determining adsorption capacity and quantity of the sludge based activated carbon which usually goes for chemical adsorption. The sludge based adsorbents are produced by anaerobic pyrolysis under approximately 900 °C which has higher adsorption capacity than commercial ones to remove metals like lead, zinc, copper and cadmium [55].

In case of high pH, the heavy metals convert to hydroxide and precipitate on the surface of sludge based adsorbents and in case of low pH, surface precipitation is less and many heavy metal ions gets exchanged with calcium ions and further adsorbed on the surface of those adsorbents. As compared to the BET and micropore volume ratio of coir and coal, the sludge based activated carbons have smaller value but the equilibrium adsorption efficiencies of lead (II), cadmium (II), chromium (VI) and copper (II) in case of sludge based adsorbents are quite higher than the commercial ones due to high acid group content. Many a times the comparative study between adsorption effect on copper (II) and lead (II) removal with the help of sludge based activated carbons activated with zinc chloride and that of commercial coal carbon were studied [56]. Although the results implied that the pore volume and BET surface area of sludge based adsorbents activated was lesser than that of the commercial ones but due to the presence of acid functional group on the surface its equivalent adsorbate uptake on the two metals was much higher than the commercial ones.

There are some heavy metal ions that are sedimented on the sludge based activated carbon's surface and they are being removed through physical adsorption which has increased adsorption capacity [57, 58]. From the research data it has been found that the adsorbent prepared with sludge and bagasse through pyrolysis under 800 °C for 0.5 h and further treatment with 60% nitric acid yielded product with around 806.57 m^2/g BET surface area [59]. As per the studies, when the sludge based activated carbon was loaded with nano-titanium oxide using the impregnation sintering method to remove the mercury ion, the performance of adsorption and efficiency of catalysis were high with mercury ion removal rate from 20 mg/L aqueous solution was around 88.5.

5.3 *Gas Pollutant Removal*

The surface of sludge based activated carbon contains a certain amount of active components whose functional groups are rich which helps to contact with the reaction gas [60]. As per the research data, the sludge based adsorbent contains a large number of micropores, ultramicropore and other nitrogen containing group which helps in low concentration formaldehyde adsorption from air [61]. The optimum adsorption rate of this sludge based activated carbon is around 83% which is nearly equal to that of commercial activated carbon. The studies have shown that the activated carbon from sludge which contains nitric acid iron sludge based catalyst facilitates a maximum 98.3% conversion of gaseous oxides of nitrogen [62]. Similarly, if titanium oxide photocatalyst is used, then photocatalytic degradation of acetone gas yields great results [63]. Also hydrogen sulphite gas was removed efficiently using sludge based activated carbon mixed with an active agent zinc chloride which was further improved with Cerium [64]. Further, phosphoric acid was also mixed with sludge to react chemically and yield mesoporous activated carbons with surface area of about 300 m^2/g [65]. The adsorption capacity of sulfur dioxide gas is associated with the average size of micropore and can controlled by the ratio of impregnation that is used mainly to make the activated carbons.

5.4 *Others*

The sludge-based adsorbents also can be combined with other water treatment processes to reduce the operating cost of various other processes. If we couple these adsorbents with membrane bioreactors for treating waste leachate then the structure and properties of the cake layer on the surface of the membrane can get better resulting in good performance of filtration and water permeability. The merits of this combination process are it reduces membrane fouling, protein and humic acid, increases the duration of membrane operation cycle and decreases the operation cost. During liquefaction of sludge based activated carbon, the energy density and yield of bio-oil at around 350 °C usually increases [66]. It is generally denoted as 350-SBAC. It helped in lowering the risk of copper, lead, cadmium and zinc. On the other hand, the sludge based activated carbon liquefied at temperature around 400 °C favored the risk reduction of strontium more. In terms of the yield of bio-oil, liquefaction is done at 350 °C with SSAC-550 was more suitable.

6 Conclusion

A generous amount of activated sludge is produced from sewage treatment. Conversion of sludge into activated carbon can bring considerable economic value and reduce environmental pollution. As compared to the traditional activated carbon, the cost of production of sludge based activated carbon is lesser because of the availability of its wide range of source. Therefore, there is a great potential value associated with the research and application of sludge based activated carbon and also presently it has obtained a certain achievements. However, some problems still remain to be unsolved and hence require further processing. Firstly, potential release of some hazardous and toxic substances during the synthesis of sludge based activated carbon was noticed. For example, it is possible to release the heavy metals from sludge based activated carbon. The mechanism of conversion of soluble heavy metal into insoluble metal compound is still not clear in the synthesis of sludge based activated carbon. Secondly, the effects of sludge based activated carbon on environment need to be further studied such as the effective reuse and recycle of sludge based activated carbon adsorption materials, the discarding of waste sludge based activated carbon, the leakage of its adsorbed substance in the transfer, the renewal technology and regeneration performance comparison between commercial activated carbon and sludge based activated carbon. Thirdly, profound study is required on the reaction mechanisms of preparation of sludge based activated carbon. Due to the complexity of sludge composition and the influence including factors like pyrolysis equipment and pyrolysis conditions etc., in the synthesis process of activation, the organic matters in activated sludge can initiate chemical reaction as a result of the activation by temperature. In the interim, the additives and chemical activators complicate chemical reaction even more. Therefore, analyzing the variations of the activation mechanism and process of activation can guide in the synthesis, application and variation of sludge based activated carbon at a broader level (Fig. 1).

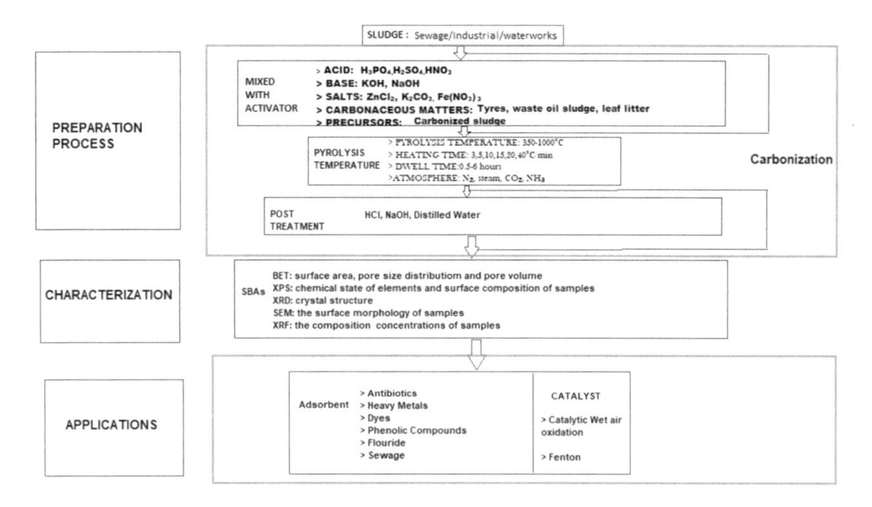

Fig. 1 Process flowsheet for the preparation, characterization and application of sludge based adsorbents

Acknowledgements The authors gratefully acknowledge the editors Prof. Eric Lichtfouse and Mr. Ali Khadir for granting this wonderful opportunity of contributing a manuscript in this highly esteemed edition "Inorganic-Organic Composites for Water and Wastewater Treatment". The authors would also like to thank the entire Department of Chemical Engineering, Sathyabama Institute of Science and Technology for facilitating all the necessary help and support throughout this journey. And lastly the authors would like to thank the authors of referred papers for providing the valuable information in them.

References

1. Edris G, Alalayah WA (2017) Sludge production from municipal wastewater treatment in sewage treatment plant. AyhanDemirbas
2. Chu W (2001) Dye removal from textile dye wastewater using recycled alum sludge
3. Bian Y, Yuan Q, Zhu G, Ren B, Hursthouse A, Zhang P (2018) Recycling of waste sludge: preparation and application of sludge-based activated carbon
4. Xu G, Yang X, Spinosa L (2014) Development of sludge-based adsorbents: preparation, characterization, utilization and its feasibility assessment
5. Otero M, Rozada F, Calvo LF, Garcı´a AI, Mora´n A (2003) Elimination of organic water pollutants using adsorbents obtained from sewage sludge
6. Hossain MK, Strezov V, Chan KY, Ziolkowski A, Nelson PF (2011) Influence of pyrolysis temperature on production and nutrient properties of wastewater sludge biochar. J Environ Manag 92(1):223–228
7. Sanchez (2009) Effect of pyrolysis temperature on the composition of the oils obtained from sewage sludge. Biomass Bioenergy 33(6–7):933–940
8. Zhang B, Xiong S, Xiao B, Yu D, Jia X (2011) Mechanism of wet sewage sludge pyrolysis in a tubular furnace. Int J Hydrogen Energy 36(1):355–363
9. Mendez A, Fidalgo JM, Guerrero F, Gasco G (2009) Characterization and pyrolysis behavior of different paper mill waste materials. J Anal Appl Pyrol 86(1), 66–73; Mendez A, Barriga S, Fidalgo JM, Gasco G (2009) Adsorbent materials frompaper industry waste materials and their use in Cu (II) removal from water. J Hazard Mater 165(1–3):736–743
10. Mahapatra K, Ramteke DS, Paliwal LJ (2012) Production of activated carbon fromsludge of food processing industry under controlled pyrolysis and its applicationfor methylene blue removal. J Anal Appl Pyrolysis 95:79–86
11. Kante K, Qiu J, Zhao Z, Cheng Y, Bandosz TJ (2008) Development of surfaceporosity and catalytic activity in metal sludge/waste oil derived adsorbents:effect of heat treatment. Chem Eng J 138(1–3):155–165
12. Yilmaz AE, Boncukcuoglu (2011) Waste utilization: the removal of textile dye (Bomaplex Red CR-L) from aqueous solution onsludge waste from electrocoagulation as adsorbent. Desalination 277(1–3):156–163
13. Ding R, Zhang P, Seredych M, Bandosz TJ (2012) Removal of antibiotics from water using sewage sludge and waste oil sludge-derived adsorbents. WaterRes 46(13):4081–4090
14. Seredych M, Bandosz TJ (2007) Sewage sludge as a single precursor for developmentof composite adsorbents/catalysts. Chem Eng J 128(1):59–67
15. Liu J, Jiang X, Zhou L, Han X, Cui Z (2009) Pyrolysis treatment of oil sludge andmodel-free kinetics analysis. J Hazard Mater 161(2–3):1208–1215
16. Velghe RC Yperman J (2012) Characterization of adsorbents prepared by pyrolysis of sludge and sludge/disposal filter cake mix
17. Ren X, Liang B, Liu M, Xu X, Cui M (2012) Effects of pyrolysis temperature, timeand leaf litter and powder coal ash addition on sludge-derived adsorbents fornitrogen oxide. Biores Technol 125:300–304

18. Ncibi MC, Jeanne-Rose V, Mahjoub B, Jean-Marius C, Lambert J, Ehrhardt JJ, Bercion Y, Seffen M, Gaspard S (2009) Preparation and characterization ofraw chars and physically activated carbons derived from marine Posidoniaoceanica (L.) fibres. J Hazard Mater 165(1–3):240–249
19. Foo KY, Hameed BH (2009) A short review of activated carbon assisted electrosorptionprocess: an overview, current stage and future prospects. J Hazard Mater 170(2–3):552–559
20. Xin-hui D, Srinivasakannan C, Jin-hui P, Li-bo Z, Zheng-yong Z (2011) Comparison of activated carbon prepared from Jatropha hull by conventional heating and microwave heating. Biomass Bioenergy 35(9):3920–3926
21. Jindarom C, Meeyoo V, Kitiyanan B, Rirksomboon T, Rangsunvigit P (2007) Surface characterization and dye adsorptive capacities of char obtained frompyrolysis/gasification of sewage sludge. Chem Eng J 133(1–3):239–246
22. Hofman M, Pietrzak R (2012) NO2 removal by adsorbents prepared from wastepaper sludge. Chem Eng J 183:278–283
23. Smith KM, Fowler GD, Pullket S, Graham NJD (2009) Sewage sludge-basedadsorbents: a review of their production, properties and use in water treatmentapplications. Water Res 43(10):2569–2594
24. Lillo-Rodenas (2008) Further insights into the activation process of sewage sludge based precursors by alkaline hydroxides. Chem Eng J 142(2):168–174
25. Monsalvo VM, Mohedano AF, Rodriguez JJ (2011) Activated carbons from sewage sludge: application to aqueous-phase adsorption of 4-chlorophenol.Desalination 277(1–3):377–382
26. Marsh H, Rodríguez-Reinoso F (2006) Activated carbon. Oxford, pp 322–365
27. Tsai J, Chiang H, Huang G, Chiang H (2008) Adsorption characteristics of acetone, chloroform and acetonitrile on sludge-derived adsorbent, commercial granularactivated carbon and activated carbon fibers. J Hazard Mater 154(1–3):1183–1191
28. Lin QH, Cheng H, Chen GY (2012) Preparation and characterization of carbonaceousadsorbents from sewage sludge using a pilot-scale microwave heatingequipment. J Anal Appl Pyrolysis 93:113–119
29. Hwang H, Choi W, Kim T, Kim J, Oh K (2008) The preparation of an adsorbentfrom mixtures of sewage sludge and coal-tar pitch using an alkaline hydroxideactivation agent. J Anal Appl Pyrol 83(2):220–226
30. Ocampo-Perez R, Rivera-Utrilla J, Gomez-Pacheco C, Sanchez-Polo M, Lopez-Pe~nalver JJ (2012) Kinetic study of tetracycline adsorption on sludge-derivedadsorbents in aqueous phase. Chem Eng J 213:88–96
31. Stüber F, Smith KM, Mendoza MB, Marques RRN, Fabregat A, Bengoa C, Font J, Fortuny A, Pullket S, Fowler GD, Graham NJD (2011) Sewage sludge based carbons for catalytic wet air oxidation of phenolic compounds in batch and trickle bed reactors. Appl Catal B Environ 110:81–89
32. Gomez-Pacheco CV, Rivera-Utrilla J, Sanchez-Polo M, Lopez-Pe~nalver JJ (2012) Optimization of the preparation process of biological sludge adsorbents for application in water treatment. J Hazard Mater 217–218:76–84
33. Lebigue J et al (2010) Application of sludge-based carbonaceous materials in a hybrid water treatment process based on adsorptionand catalytic wet air oxidation
34. Zou et al (2013) Structure and adsorption properties of sewage sludge-derived carbon with removal of inorganic impurities and high porosity
35. Xuemin H, Xin S, Quan Y (2013) Adsorption of toluene by fixed sludge-based activated carbon bed. Chin J Environ Eng 7:1085–1090
36. Ping F, Ruihua S, Juan R (2011) Adsorption of phenol from aqueous solution using activated carbon. Carbon Techniq 30:12–16
37. Daojing L (2011) Study on sludge activated carbon preparation and its adsorption properties of phenol and nitrobenzene. Beijing Forestry University
38. Lijun Y, Wenju J (2005) Adsorption of red-water from trinitrotoluene manufacturing by sludge-based adsorbent. Ind Water Wastewater 36:26–28

39. Jiarong H (2016) The adsorption study of rhodamine-B by activated carbon of activated sludge. J Minnan Normal Univ (Natural Science) 83–87
40. Xue W, Liqiu Z, Li F (2015) Removal efficiency of ibuprofen and determination of active sites in catalytic ozonation process by modified SCACs. Chin J Environ Eng 9:621–626
41. Zhihui P, Chaosheng Z, Jiayu T, Qing Z, Guibai L (2014) Application of chemical sludge based adsorbent in wastewater treatment. Water Wastewater Eng 40:142–145
42. Xin S, Xue-min H, Li C, Quan Y (2012) Preparation of sludge-based activated carbon and adsorptive properties of toluene. Environ Sci Technol 35:32–35
43. Hongjuan W, Fei Q, Li F, Liqiu Z (2012) Catalytic ozonation of ibuprofen in aqueous solution by activated carbon made from sludge and corn cob. Environ Sci 33:1591–1596
44. Daojing L, Li F, Liqiu Z (2012) Effects of corncob addition onproperties of sludge activated carbon. Chin J Environ Eng 6:1010–1014
45. Wen G, Pan Z-H, Ma J, Liu Z-Q, Zhao L, Li J-J (2012) Reuse of sewage sludge as a catalyst in ozonation—efficiency for the removal of oxalic acid and the control of bromated formation. J Hazard Mater 239–240:381–388
46. Zhuang H, Han H, Hou B, Jia S, Zhao Q (2014) Heterogeneous catalytic ozonation of biologically pretreated Lurgi coal gasification wastewater using sewage sludge based activated carbon supported manganese and ferric oxides as catalysts. Biores Technol 166:178–186
47. Huang Y, Sun Y, Xu Z, Luo M, Zhu C, Li L (2017) Removal of aqueous oxalic acid by heterogeneous catalytic ozonation with MnOx/sewage sludge-derived activated carbon as catalysts. Sci Total Environ 575:50–57
48. Haifeng Z (2015) Research on efficiency of advanced treatment of coal gasification wastewater by catalytic ozonation integrated with biological process. Harbin Institute of Technology
49. Yangyang Y, Xueqiang L, Danyu X, Tao Z, Yan S, Ang Y (2015) The catalytic ozonation of sludge-based composite activated carbon for the degradation of Rh B in aqueous solution. Ind Water Treatment 35:56–59
50. Gu L, Zhu N, Zhou P (2012) Preparation of sludge derivedmagnetic porous carbon and their application in Fenton likedegradation of 1-diazo-2-naphthol-4-sulfonic acid. Biores Technol 118:638–642
51. Sun H, Peng X, Zhang S et al (2017) Activation of peroxymonosulfate by nitrogen-functionalized sludge carbon for efficient degradation of organic pollutants in water. Bioresour Technol 241:244–251
52. Gupta A, Garg A (2015) Primary sewage sludge-derived activated carbon: characterization and application in waste water treatment. Clean Technol Environ Policy 17(6):1619–1631
53. Kong L, Xiong Y, Sun L et al (2014) Sorption performanceand mechanism of a sludge-derived char as porous carbonbasedhybrid adsorbent for benzene derivatives in aqueoussolution. J Hazard Mater 274:205–211
54. Nunthaprechachan T, Pengpanich S, Hunsom M (2013) Adsorptive desulfurization of dibenzothiophene by sewagesludge-derived activated carbon. Chem Eng J 228:263–271
55. Tan C, Rong H, Hongtao W, Wenjing L, Yuancheng Z, Zeyu Z (2014) Adsorption of heavy metals by biochar derived from municipal sewage sludge. J Tsinghua Univ 54:1062–1067
56. Weiwei Y, Li F, Liqiu Z (2014) Preparation of columnarsludge-based activated carbon and its application in pollutantsremoval. Acta Sci Circum 34:385–391
57. Ziyan F (2014) Technological research of adsorption of heavy metals by modified activated carbon. Tsinghua University
58. Zaini MAA, Zakaria M, Alias N et al (2014) Removal of heavymetals onto KOH-activated ash-rich sludge adsorbent. Energy Procedia 61:2572–2575
59. Tao HC, Zhang HR, Li JB, Ding WY (2015) Biomass based activated carbon obtained from sludge and sugarcane bagasse for removing lead ion from wastewater. Bioresour Technol 192:611–617
60. Yifan Z, Kangsheng B (2017) Study on the adsorption characteristicof formaldehyde in the air by modified PSAC. AnHuiChem Ind 43:33–37
61. Qingbo W, Caiting L, Zhihong C, Wei Z, Hongliang G (2010) Application of sewage sludge based activated carbon inform aldehyde adsorption. China Environ Sci 30:727–732

62. Tao L (2007) Preparation and properties of nitrogen oxide catalyst derived from sewage sludge. Hunan University
63. Yanjing Z (2015) Study on preparation of TiO_2/sludge activated carbon and its photocatalytic purification of acetone gas. Hebei University of Science & Technology
64. Wei J, Linhuan Z, Fen L, Bo Y, Anxi J (2016) Study on preparationand performance of cerium modified active carbonadsorbent from sewage sludge. Mater Rev 30:411–414
65. Boualem T, Debab A, Martínez de Yuso A, Izquierdo MT (2014) Activated carbons obtained from sewage sludgeby chemical activation: gas-phase environmental applications. J Environ Manage 140:145–151
66. Zhai Y, Chen H, Xu B et al (2014) Influence of sewage sludgebasedactivated carbon and temperature on the liquefactionof sewage sludge: yield and composition of bio-oil, immobilizationand risk assessment of heavy metals. Bioresour Technol 159:72–79
67. Al-Malack MH, Dauda M (2017) Competitive adsorption ofcadmium and phenol on activated carbon produced from municipal sludge. J Environ Chem Eng 5(3):2718–2729
68. Hanfeng B (2013) Performance and mechanism of four kinds of heavy metals removal from water by prepared sludge-based activated carbon. Beijing Forestry University
69. Qing L, Xueying R, Jiali L (2012) The Preparation of titanium dioxide photocatalyst loaded on the modified products of municipal sewage sludge. Guangdong Chem Ind 39:59–60
70. Dezhi L, Guangzhi W, Xin L, Ping W (2015) Study on membrane fouling properties in treatment of landfill leachate by SBAC/MBR process. China Water Wastewater 31:21–26
71. Razali M, Zhao YQ, Bruen M (2006) Effectiveness of a drinking water treatment sludge in removing different phosphorus species from aqueous solution
72. Book named 3—Municipal wastewater and sludge treatment
73. Andreoli CV, von Sperling M, Fernandes Book named biological wastewater treatment series, vol 6. Sludge treatment and disposal

Printed in the United States
by Baker & Taylor Publisher Services